Suelem Vieira dos Santos
Wagner Cherobin

Hybrid solar collector

Suelem Vieira dos Santos
Wagner Cherobin

Hybrid solar collector

Hybrid solar collector for energy generation

ScienciaScripts

Imprint
Any brand names and product names mentioned in this book are subject to trademark, brand or patent protection and are trademarks or registered trademarks of their respective holders. The use of brand names, product names, common names, trade names, product descriptions etc. even without a particular marking in this work is in no way to be construed to mean that such names may be regarded as unrestricted in respect of trademark and brand protection legislation and could thus be used by anyone.

Cover image: www.ingimage.com

This book is a translation from the original published under ISBN 978-613-9-70195-7.

Publisher:
Sciencia Scripts
is a trademark of
Dodo Books Indian Ocean Ltd. and OmniScriptum S.R.L publishing group

120 High Road, East Finchley, London, N2 9ED, United Kingdom
Str. Armeneasca 28/1, office 1, Chisinau MD-2012, Republic of Moldova, Europe
Printed at: see last page
ISBN: 978-620-8-17554-2

ACKNOWLEDGEMENTS

To God, for giving us the privilege of life and allowing us to complete another stage of our journey.

We would like to thank our families for their support, encouragement and understanding of the times we were absent to complete this work.

We would like to thank the company Imetal for their collaboration and support in completing this work, especially Everton for his understanding during times of absence.

We would like to thank our teacher and supervisor, Joiris Manoela Dachery, for her commitment and special dedication to this work.

A very special thanks to Professor and engineer Marcellus for his great collaboration and support in bringing this work to a successful conclusion.

We would also like to thank Professor Gilson, who with his great knowledge guided us in the right direction for our final work.

To my friends and colleagues, especially Leti and Thais, for the moments of distraction, the exchange of experience and all the beautiful moments we spent together that will stay with us forever.

To everyone who helped in one way or another to make this work a success, thank you very much!

"Maybe I didn't manage to do the best, but I fought for the best. I'm not what I should be, but thank God I'm not what I was before". (Marthin Luther King)

SUMMARY

Solar energy is the most abundant of renewable energies, as it is the first and main energy that gives rise to life. This project aims to compare how photovoltaic panels work, demonstrating their construction and studying the effect of cooling on a PV module, as well as analysing the use of solar energy in a single module, a hybrid solar collector (photovoltaic and thermal). The concept of converting solar energy into electricity and heat was explained. In photovoltaic generation, performance is directly linked to operating conditions such as incident solar radiation, construction model and operating temperature. With the construction and assembly of the experiment, data such as voltage and current were analysed, as well as the temperatures involved in the system. The data for the experiment was obtained between 29/03/2016 and 19/05/2016. The measurements were taken from 12:00 pm to 6:00 pm. The PV module with a thermal collector was characterised as CT and the ST module without a thermal collector, i.e. conventional. The electrical generation efficiency of the CT module averaged 4.55%, while the ST module had an average efficiency of 4.22%. The CT module generated the most energy during the days it was collected, totalling 569.3 W, while the ST module generated 536.5 W, with a difference of 32.8 W. It can be concluded that the cooling of an amorphous silicon PV panel tends to generate more electricity than a conventional PV module.
Keywords: Solar energy. Efficiency. Photovoltaic. Thermal. Hybrid.

SUMMARY

CHAPTER 1

INTRODUCTION

The idea of combining photovoltaic (PV) and solar thermal panels is not new; there is already research and companies in the sector adhering to the new hybrid solar collector model. The main objective of a hybrid solar collector (thermal/PV) is to simultaneously produce electricity and heat, says Almeida (2008). The main technology applied in the use of a hybrid collector lies in controlling the temperature of the photovoltaic cells, increasing the efficiency of electricity generation and also extending their useful life.

At first glance, the operation of a solar thermal collector is simple and based on heat transfer phenomena, energy from solar radiation to heat a fluid. When solar radiation hits the surface of the panel, some is absorbed, some is reflected and the rest is transmitted. The direct conversion of solar energy into electrical energy occurs through the effects of solar radiation on certain semiconductor materials, which is known as the photoelectric effect (TIRADENTES, 2007).

Atmospheric conditions such as cloudiness and the amount of solar radiation are directly dependent on the local latitude and the tilt of the Earth's axis, providing a longer or shorter duration of sunlight or clarity. Due to the fact that much of Brazil's territory is close to the equator, our country is potentially a major producer of energy using solar radiation (RÚTHER, 2004).

According to the National Electric Energy Agency, ANEEL (2002), the highest radiation levels are observed in the Northeast, where the annual average is approximately 6.0 kW/m^2 .day. The lowest rates are observed on the South-Southeast Coast, including the Serra do Mar, and in Western Amazonia. In Amapá and eastern Pará, rates are also lower than the national average. It is important to emphasise that even the regions with the lowest radiation rates have great potential for energy use.

Currently there are several installations of photovoltaic panels that are interconnected to the conventional electricity grid, injecting the energy produced during the day into the grid and at night the user uses the energy needed for consumption from the conventional grid (CÂMARA, 2011).

1.1. GENERAL OBJECTIVES

The aim of this work was to build a hybrid solar collector model and compare the two ways of harnessing solar energy in a single panel, photovoltaic and thermal.

1.2. SPECIFIC OBJECTIVES

-Study how hybrid solar collectors work;

-Build and assemble a model of a hybrid solar collector;

-Make a comparison between a PV solar panel and a hybrid collector;

-Analyse and interpret the data obtained in the experiment;

-Evaluate temperatures and their influence on the system.

CHAPTER 2

THEORETICAL BACKGROUND

2.1. ENERGY

Energy is directly related to life and is present in all of nature's activities. As humanity evolves, the demand for energy is increasing, which is why it is necessary to look for alternative energy sources. Electricity has become one of the most variable and useful forms of energy. It has therefore become an indispensable and strategic resource for the socio-economic development of many countries and regions (NUNES, 2011).

The production of electricity was and is important for development in the modern world. The discovery of electricity made it possible to improve health, water supply and food supply, quality of life and sanitary conditions. It also contributed to establishing the capitalist system and consumer society (GUENA, 2010).

2.1.1. Renewable Energies

Renewable energy is considered to be natural resources that are capable of regenerating themselves, unlike non-renewable sources such as oil. The best known renewable energy sources are: wind, solar, biomass, biofuel, biogas, geothermal, hydroelectric and sea energy, among others (FILHO & JÚNIOR, 2007).

2.1.2. Solar Energy

The Sun is the primary source of most of the energy that exists today. It is the electromagnetic radiation incident by the Sun on the Earth that makes it possible to produce heat and power (MATIAS, 2014).

According to ANEEL (2010), the most widely used processes for harnessing solar energy are water heating and photovoltaic electricity generation. In Brazil, water heating is more common in the South and Southeast, due to climatic characteristics, and photovoltaic generation in the North and Northeast, in communities isolated from the electricity grid.

According to Lopez (2012), let's look at some information about solar energy:

- The solar energy incident on the Earth each day is equivalent to all the energy consumed in the world for 27 years;
- The amount of solar radiation on Earth every three days is equivalent to the energy stored in

all the world's known fossil energy sources (oil, natural gas, coal);

- The transmission of solar energy to the Earth occurs through electromagnetic radiation, 97% of which is contained between wavelengths ranging from 0.3 to 3.0 (micrometres), characterising shortwave radiation;
- Under clear skies and without clouds, the maximum radiation observed at midday at a location at sea level is 1.0 kW/m^2 at a height of 1,000 m, the radiation is 1.05 kW/m ;^{2}
- Brazil has an excellent solar radiation index, especially in the northeast, with typical values ranging from 1,752 kWh/m^2 to 2,600 kWh/m^2 per year.

2.2. HEATING SYSTEMS

As a renewable energy, solar energy can be harnessed in a variety of ways, thermal energy being one of them. According to Tiradentes (2007), heating is the most natural and obvious application of solar energy, used by man since prehistoric times. And over the centuries, various ways of improving its utilisation have been developed, albeit in a primitive or rustic manner. History mentions the famous naval battle of Syracuse, where Archimedes, using curved mirrors as a way of concentrating the sun's light and heat, set fire to the Roman fleet besieging the city. This happened around 215 BC.

Various applications of solar energy for heating fluids, solids or environments are now well known. In the coldest regions of the planet, it has developed remarkably, with the use of simple but efficient technologies, at least during the periods of greatest insolation.

According to ABRAVA (2008), solar water heaters also contribute to economic development in various ways. For example, since they do not require large capital investments for their production, they can be produced by small and medium-sized companies, which are recognised as important job generators. The decentralisation intrinsic to this technology is also responsible for significant job creation in retail, design and installation companies.

Teixeira (2009), in his master's thesis, says that the aim of solar thermal collectors is to convert the greatest amount of available solar radiation into heat and transfer this heat with as few losses as possible to the rest of the system. There are various types and designs of collectors with different costs and performances for a wide variety of applications.

The main problem with utilising solar energy lies in the almost total dependence on sunny days for installed systems to perform well. The conversion of solar energy into heat does not allow it to be stored for long periods, in practice it is a form of energy that must be utilised almost immediately (TIRADENTES, 2007).

The author Paim (1994, p. 112) apud Tiradentes (2007) emphasises that the performance of a heat collector will always depend on its location. In certain regions, the use of some systems can become impractical due to climatic and geographical influences. In any case, the use of solar collectors requires careful sizing, taking into account the minor climatic conditions of the user's location. This requires precise measurements of the solar intensity profile over days, weeks and years.

2.2.1. Types of heating systems

Harnessing the energy of solar radiation requires devices with surfaces capable of absorbing as much of the incident solar radiation as possible and transferring this absorbed energy to the fluid to be used (RORIZ, 2008).

Roriz (2008) says that a solar collector is a device that must have a surface with high absorptivity in order to maximise the capture of incident radiant energy, be thermally insulated in order to reduce thermal losses through convection and conduction, and have a low emissivity for the wavelengths of its own radiation.

The main types of thermal collector systems are classified according to how the thermal fluid is circulated, namely thermosiphon circulation (passive) and forced circulation (active).

2.2.2 Thermosiphon circulation (passive)

In the thermosiphon system, circulation through the solar collectors occurs due to the difference in specific weight of the thermal fluid in the regions of the accumulation reservoir and the collectors. The fluid heated in the collectors becomes lighter and tends to move upwards by the principle of convection, while the fluid in the reservoir region, due to being colder and therefore heavier, tends to move downwards towards the collectors, thus establishing a circulation process that continues as long as there is a temperature difference (GUARIENTE, 2005).

This operating characteristic requires rigour in the positioning of the accumulation tank and the collectors. The tank must always be at a level above the collectors, as in Figure 1, so that it is possible for the heated fluid to move from a lower level to a higher level and to hinder reverse flow, which can occur at night due to the inversion of temperatures between collectors and tanks (Rudnick et aL, 1986, apud Guariente, 2005).

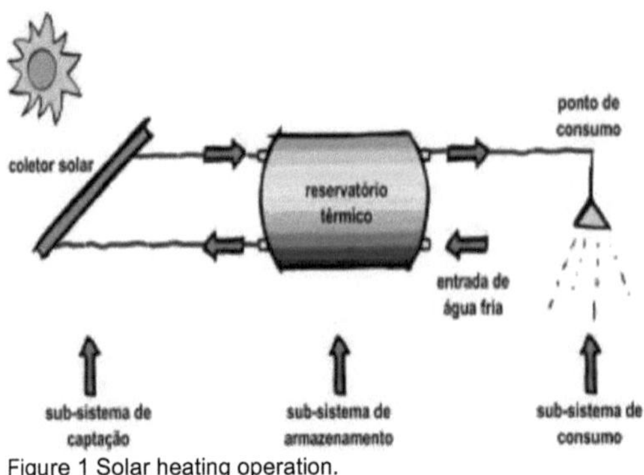

Figure 1 Solar heating operation.
Source: Pucminas, 2015.

2.2.3. Forced circulation (asset)

In the active system, according to Guariente (2005), the circulation of the thermal fluid between the collectors and the accumulation reservoir is carried out with the aid of a hydraulic pump that transmits energy to the process. The use of auxiliary energy allows for greater versatility in the positioning of the components, and the reservoir can be placed at any level in relation to the collectors, even below them, as shown in Figure 2.

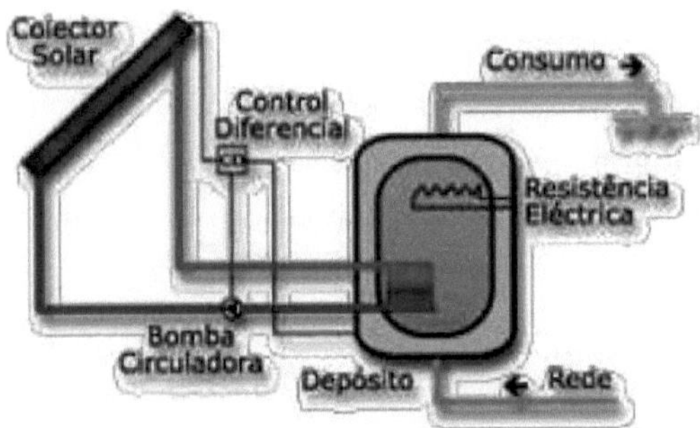

Figure 2 Forced circulation scheme.
Source: Solaribérico, 2015.

The pump can be activated using thermostats that control the system's temperature, since promoting the circulation of the thermal fluid in unfavourable conditions can cause the water to lose heat rather than heat up. Timers can also be used to activate the pump at specific times when the incidence of solar radiation is more favourable and greater use can be made of the energy (GUARIENTE, 2005).

In Brazil, almost all of the residential solar water heating systems used are of the direct passive type. In some cases, due to architectural requirements or because they are large systems, it is necessary to opt for a direct active system (SIQUEIRA, 1996, apud GUARIENTE, 2005).

2.2.4 Flat plate thermal collector

There are various types of solar collectors, such as vacuum tubes and flat plates. This work will use the flat plate type. It consists mainly of an absorber plate, coils in tubes, insulators and a reservoir. In this type of collector, the heat is transferred to water, which circulates through the pipework in the circuit, which in turn transfers the heat to a reservoir (DUARTE, 2012).

Petrucci (1998, p.18) apud Guariente (2005), describes the constitution of a thermal reservoir as cylindrical in shape, made from stainless steel or carbon steel sheet metal with an internal coating of anti-corrosive material. A material with low thermal conductivity is used to thermally insulate the internal tank. Glass wool, polyurethane foam, etc. can be used as thermal insulation. Another external metal sheet is used to close and support the insulation.

2.3 PHOTOVOLTAIC CELLS

The first solar cells were developed in the 1950s for use in artificial satellites. The efficiency of these cells is still low today, at around 18 per cent, because most of the radiant energy collected is lost as heat (TIRADENTES, 2007).

Photovoltaic cells are mostly made from monocrystalline or polycrystalline silicon. With different areas and thicknesses. The standard test or reference conditions are set at around 1000 W/m^2 , at a cell temperature of 25 °C, generating a current of approximately 1.5 A (50 cm^2) and 4.5 A (150 cm^2) at a voltage of 0.46 V and 0.48 V.

Figure 3 Detail of a photovoltaic cell
Source: Kugler (2010).

To provide electrical insulation, a transparent polymer, usually ethyl vinyl acetate (EVA), is placed over the cells. In addition, photovoltaic modules have tempered, anti-reflective glass for the front cover, polyvinyl fluoride on the back and an electrical junction box also on the back. These cells together form a PV module, and by organising several modules we get a PV arrangement, shown in Figure 4 (PRIEB, 2002).

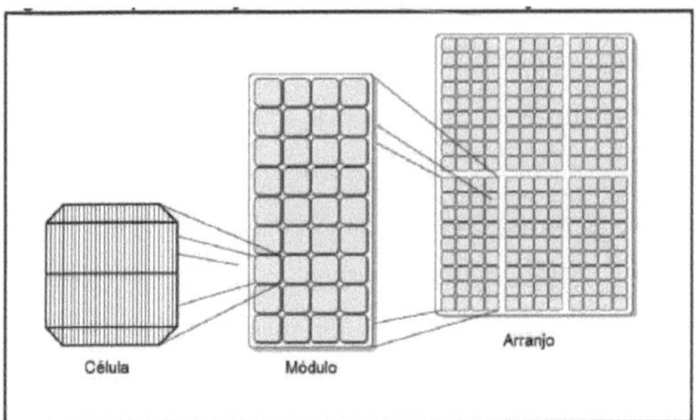

Figure 4 Representation of photovoltaic cell, module and arrangement
Source: Rogani (2013).

Amorphous silicon (a-Si) photovoltaic modules, on the other hand, have different construction characteristics. They basically consist of several layers of deposited material. To begin with, the a-Si module is deposited on TCO (Transparent Conductive Oxide) glass. Then layers of impurity-laden semiconductor material with an excess of positive charges, known as the "p" type, are deposited, as well as a layer of intrinsic material that divides the layers and another layer of negatively charged semiconductor material, known as the "n" type layer.

According to the author Guimarães (2010), a layer of ZnO doped with aluminium is also deposited to improve the module's electrical contact. Another layer of glass is added to close the layers and assemble the module on its support structure, as shown in Figure 5.

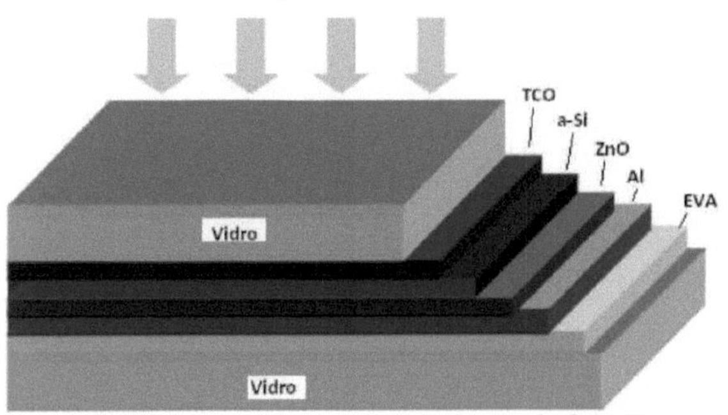

Figure 5 Demonstration of the manufacturing layers of a Solar Plus module.
Source: Guimarães (2010).

2.3.1 Photovoltaic conversion

According to Zilles et al (2012, p. 13), the conversion of the energy contained in light radiation into electrical energy is a physical phenomenon known as the photovoltaic effect, in which certain semiconductor materials are able to absorb the energy contained in the photons present in the incident light radiation, transforming it into electricity. The energy absorbed by these materials breaks down their structural molecules and electrical charges are released, which can be used to carry out work. Buhler (2007) summarises photovoltaic conversion as the emergence of a potential difference between two semiconductor materials with different electrical properties due to the incidence of light on the junction region between them. The vast majority of devices that make use of the photovoltaic effect are made of silicon in its various forms, mainly multicrystalline silicon.

The semiconductor elements most commonly used in photovoltaic conversion are: Monocrystalline silicon (Si), Polycrystalline silicon and amorphous silicon. Although silicon crystals have a metallic appearance, it is referred to as a semi-metal.

Búhler (2007) says that a silicon atom has four electrons in its last electron shell, and binds to other silicon atoms to form a crystal lattice, with covalent bonds and no electrons available for electronic conduction. A pure silicon crystal is practically an insulator, so little electricity passes through it. The authors Pinho and Galdino (2014 p. 104) describe that semiconductor materials, such as silicon, have energy bands that, through thermal excitation, electrons jump from the valence band to the conduction band, leaving gaps or voids in the valence band. However, this thermal excitation generates very little conductivity in the material. The incidence of photons on this material also generates electron-gap pairs. Silicon also undergoes a doping process to increase its conductivity.

Still in his work, Búhler (2007) comments that doping involves mixing a small amount of impurities into a silicon crystal. When mixed into the silicon lattice, they form gaps in the molecule and an electron in the silicon has nothing to bind to. The absence of an electron creates the effect of a positive charge, hence the name p-type.

The author Zilles et al (2012) emphasises in his book that when an electron in the valence layer is hit by a photon, it absorbs the energy needed to free it from its chemical bond, moving into the conduction band, creating an electron-gap pair. The greater the incidence of photons, the more electron-gap pairs are formed and separated by the field, causing an imbalance in the junction currents and establishing a potential difference due to the accumulation of charge carriers on opposite sides of the p-n junction (electrons in the n region and gaps in the p region). Finally, metal conductors are connected to each side of the semiconductor and interconnected between the cells, and as long as there is solar radiation falling on the semiconductor, there will be an electric current. Figure 6 shows the process of converting solar energy into electricity and its utilisation.

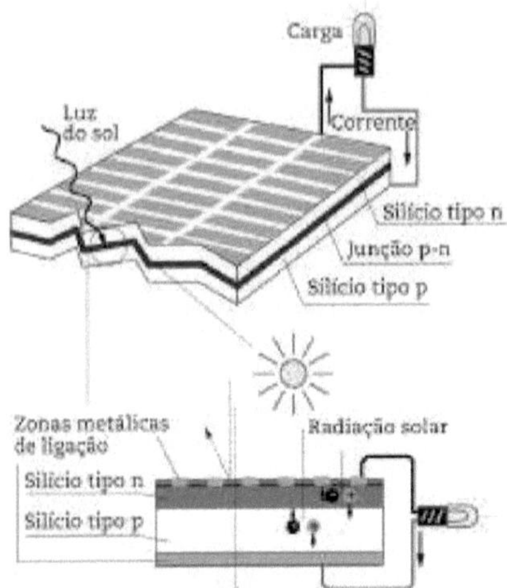

Figure 6 Representation of the photovoltaic conversion process.
Source: Zilles et al, (2012).

2.3.2 Photovoltaic panel electrical characteristics

The PV effect, as already explained, occurs when the cell is exposed to solar radiation and the energy generated is utilised through the electric field of the p-n junction. When the cell is not connected to any load, a so-called open-circuit voltage (Voc) will appear at its terminals, unlike when it is connected to a load, thus circulating current in the circuit between the load and the cell. The most important electrical characteristics of a PV module are nominal power, voltage and current. Pinho and Galdino (2014 p. 116) tell us that the electric current in a photovoltaic cell is the sum of the current of the p-n junction in the dark and the current generated by the photons absorbed from solar radiation.

The electrical circuit that describes the internal workings of a photovoltaic cell when solar radiation is emitted onto it is shown in Figure 7.

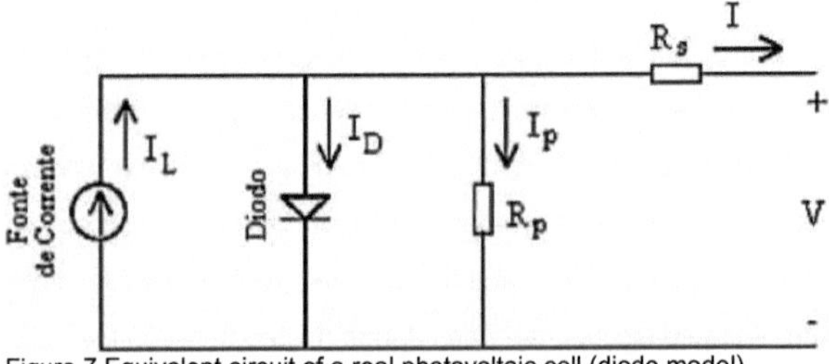

Figure 7 Equivalent circuit of a real photovoltaic cell (diode model)

Source: Hecktheuer (2001).

The parameters shown in the figure are defined as:

J Il - current generated by the incidence of radiation;

J ID - reverse saturation current of the diode;

J Ip - Earth leakage current;

J I - current at the output terminals;

J Rp - parallel resistance;

Z Rs - series resistance.

The circuit shown in Figure 7 is a real equivalent circuit, more complex and complete than the ideal equivalent circuit of a photovoltaic cell. The ideal model of a cell's electrical circuit is impossible to achieve, so some factors that alter this circuit must be taken into account. The diode "D" represents the participation of the p-n junction in the dark.

The series resistance "Rs" would be the same as the internal resistance or ohmic losses of the material to the passage of current and depends on the thickness of the junction, the impurities in the material and the contact resistance. "Rp" represents the eddy currents generated between the top and bottom of the cell, by the edge and irregularities or impurities. In an ideal cell there is no internal loss (Rs = 0) and no ground current (Rp = 0).

2.3.3 Current-voltage curve (I - V)

The curve of the photovoltaic generation graph is the useful part that most interests the professional or person who is going to install a PV system, as shown in Figure 8.

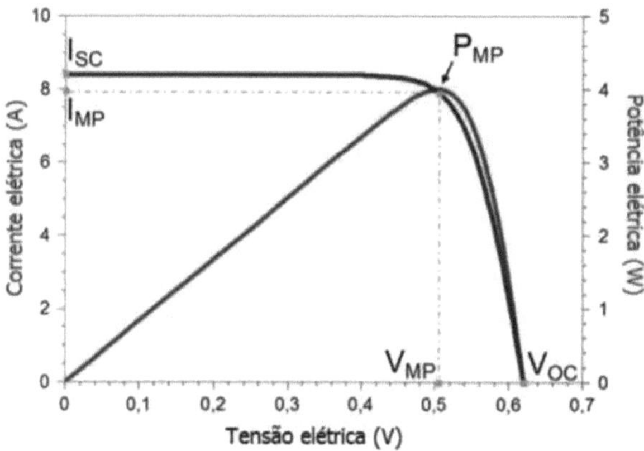

Figure 8 I-V curve with the electrical power of a photovoltaic cell.
Source: Pinho and Galdinho (2015, p. 120).

In summary of the work by Pinho and Galdino (2014 p. 119), the electrical parameters that characterise photovoltaic modules can be determined from the I-V curve. The graph above shows that the open circuit voltage (Voc) and short circuit (Isc) points do not produce any power, since power is the product of current and voltage. The Voc voltage occurs when there is no electric current circulating, so it is the maximum voltage that a photovoltaic cell can produce. The current Isc is the maximum current that can be obtained, measured when the electrical voltage at its terminals is zero.

The characteristic curves (current x voltage) represent the values of the output current as a function of voltage for a PV module. Under pre-established standard test conditions, the point of maximum power (P_{mp}) is located on the curve or "knee" of the I - V curve, indicated by the points (I_{mp} , V_{mp}). This is where the maximum transfer of power from the cell to the load occurs. Figure 8 also shows a power curve (P - V) for the same level of irradiation, which is plotted by multiplying the voltage and current values equivalent to the module's characteristic curve (I - V) point by point.

Another important observation Buhler (2007) emphasises is that the ideality factor, or shape factor, is a parameter directly associated with the curvature of the I-V curve, and therefore must be taken into account. The Form Factor (FF) defines how close the I - V curve is to the ideal. It depends very much on the construction characteristics, the type of semiconductor, doping, connections, etc. It can be defined by equation 1:

$$FF = \frac{Vmp * Imp}{Voc * Isc} \qquad Eq.1$$

According to Buhler (2007), the characteristic curve of a PV module is a mathematical representation of the relationship between electric current and voltage. The I-V curve provides information on the generator's electrical performance.

2.3.4. Factors that modify electrical characteristics

Many factors influence photovoltaic generation, some resulting from the manufacturing process and the material used, as well as the series and parallel resistances and absorption selectivity of the semiconductor. Other factors such as irradiance and cell temperature also reduce yield.

Absorption selectivity is a characteristic of the material alone, i.e. the energy level required for a photon to be absorbed by the semiconductor material. Incident radiation is the factor that most affects the performance of a PV module, and the electrical power generated is proportional to the amount of radiation incident on the plane of the cell.

The temperature of the cell during electrical generation also affects photovoltaic conversion efficiency. According to Zilles et al (2012, p. 26), this increase in temperature leads to a decrease in the open circuit voltage, automatically resulting in a decrease in the maximum power point. In his work, Guimarães (2010) mentions that the decrease in power generated for each degree of temperature of the a-Si cell above 25°C ambient temperature is 0.2%/°C. According to Andrade (2008) in his work, the influence of temperature is significant on open-circuit voltage (Voc) values, while the variation in short-circuit current values is not significantly affected. Figure 9 shows this behaviour.

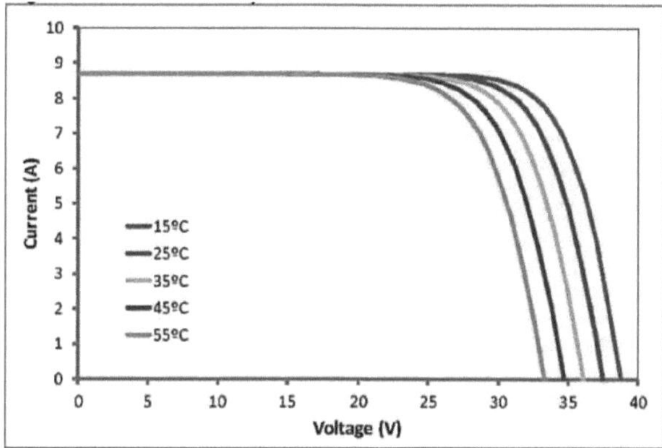

Figure 9 Influence of temperature on the current and voltage curve.
Source: Ingemecanica (2015).

2.4. HYBRID SOLAR COLLECTOR

The hybrid solar collector is a technology that converts solar radiation into electricity and heat simultaneously. It combines the functions of solar thermal and PV collectors, as explained by Charalambous et al (2004). Basically, it consists of adding a heat exchange system to a PV module.

In this way, the absorber plate or heat exchanger collects the thermal energy and cools the photovoltaic module, resulting in an improvement in its electrical behaviour. This thermal energy is then used to heat water or air. As well as achieving greater energy conversion efficiency, a hybrid system reduces the amount of usable space and initial investment compared to a system consisting of a PV module and a solar collector placed side-by-side.

Almeida (2008) found that a hybrid system with monocrystalline cells produces around 26% to 31% more energy than a conventional PV module for the same area and installed power. He also concludes that a system with a conventional collector produces around 2% to 9% more thermal energy than a hybrid collector.

2.4.1 Advantages and disadvantages of a hybrid collector

The main advantages of the hybrid collector are:

- J For roofs where the area is small, the hybrid collector excels, as it generates a greater amount of electrical and thermal energy per m^2 , than a separate solution (thermal + photovoltaic);
- J From an aesthetic point of view, the hybrid collector provides a more harmonious solution compared to combining a solar thermal collector with a photovoltaic collector;

Some disadvantages include:

- J Lower thermal efficiency compared to a conventional solar thermal collector;
- J More complicated and expensive installation, as all the plumbing and electrics will have to be installed together.
- J Total cost and return on investment in years are higher compared to conventional systems.

2.5. TEMPERATURE SENSOR

The thermocouple sensor is considered to be the most widely used of all types of thermometers as temperature pulse sockets, especially for temperature ranges up to 1000°C. The authors Nishinari and Sighieri (1966) mention that the thermocouple's working principle was discovered by Thomas Seebeck in which any temperature difference between the junctions of two different metals generates a potential difference between these junctions. The potential generated is determined by the following three factors:

-The potential difference is proportional to the temperature difference between the junctions;

-The potential depends on the combination of different metals;

-The potential depends on the homogeneity of the material;

It is also correct to say that the thermocouple sensor is the entire length of the wires between the measuring junction and the reference junction, because when the measuring junction is in contact with a heat source and the other junction is in the environment, the temperature difference between the junctions will generate a small voltage, in the millivolt range, and will be transmitted to a voltmeter already calibrated on the temperature scale. The thermocouple circuit is shown in Figure 10.

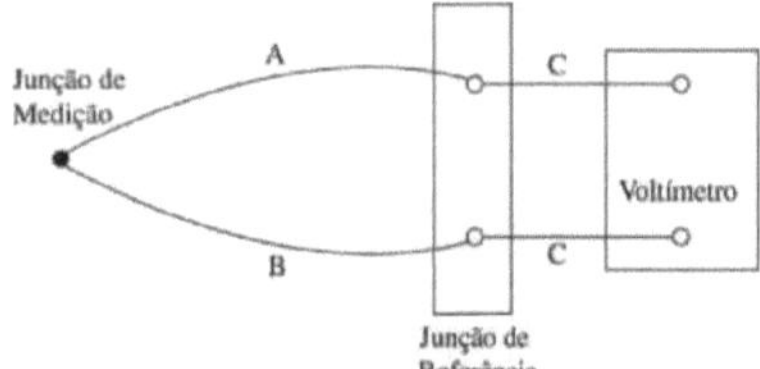

Figure 10 Circuit for measuring the Seebeck potential.
Source: Moreira, 2002.

The type K thermocouple is for general use at a lower cost. It is made up of two nickel-based metal wires called Cromei, the positive thermoelement, and Alumel, the negative thermoelement. The Chromei element contains 10% Chromium and 90% Nickel, while the Alumel element contains 94% Nickel, 3% Manganese, 2% Aluminium and 1% other. According to the thermocouple supplier's catalogue, its range of use in degrees Celsius is around -200 to 1370 °C. In the measuring range of -100 to 999 °C the accuracy is around ±0.05% (MOREIRA, 2002).

CHAPTER 3

MATERIALS AND METHODS

For this work, two photovoltaic modules were used, one of which was fitted with a heat-absorbing plate without any major alterations. This panel was labelled P1 CT (with thermal collector) and the other P2 ST (without thermal collector). The fluid flow is free with forced circulation using a small water pump. To characterise the data, some electronic equipment was needed to acquire the values generated over a certain period of time.

Based on the analyses made, the experimental results collected during the stipulated periods will be described. It was decided to collect the data on sunny, warm days from 29/03/2016 to 19/05/2016. The collection procedure took place in such a way as to utilise the highest solar incidence available. For this reason, the devices were switched on at midday and programmed to record the measurement instantaneously every 60 seconds.

3.1. INSTRUMENTATION USED

In the proposed work, voltage, current and temperature measurements were taken for the solar panel in both configurations, operating normally and using a heat exchanger. The measurements required instruments capable of measuring the electrical measurement unit and then acquiring and storing the data on a computer.

3.1.1. Photovoltaic Panels

Amorphous silicon (a-Si) photovoltaic panels were used. The panels used were built by Solar Plus, a company based in Portugal and founded in 2005, which is recognised for producing modules using thin-film technology. The construction of the a-Si modules has fragile surfaces in their structure, as they use glass on the front and back.

The photovoltaic panels used in the experiment were installed in the renewable energy laboratory in block H at UNOESC - Xanxerê campus. They are mounted on a support at a 45° inclination. Their pointing direction is north-west. This positioning of the panels was analysed according to the shading around them. The equipment mounted in the laboratory is located on a balcony on the second floor of block H. This balcony has an area of approximately 43 m^2 . Figure 11 below shows the location of the experiment.

Figure 11 Photovoltaic panel used.
Source: The authors (2016).

The electrical characteristics of the a-Si, Solar Plus photovoltaic module are listed in Table 1 below:

Table 1 Electrical characteristics of the a-Si Solar Plus photovoltaic panel.

PARAMETERS		MODEL: sp-42
Maximum power	Pmax(W)	42,0
Open circuit voltage	Voc (V)	60
Maximum peak voltage	Vmp(V)	45
Short circuit current	Isc(A)	1,18
Maximum peak current	Imp (A)	0,94
STC performance: 1000W/m irradiation², 25°C module temperature. AM 1.5, power measurement tolerance ±3%		

Source: the authors (2016).

3.1.2 Aluminium heat exchanger

The heat exchanger used in the experiment was modelled using 3D CAD software, in a size proportional to the photovoltaic module. The collector model was based on the principle of heat transmission by conduction, where the rear surface of the photovoltaic module is in contact with the surface of the thermal collector, transmitting the heat from the panel to the circulating fluid inside the heat exchanger.

The design of the heat exchanger took into account a simple design with the largest total surface area available for heat transfer.

When choosing materials, the ability to conduct thermal energy was taken into account. Thermal conductivity is an individual characteristic of each material. The collector was made from aluminium (5052 - H114) which has a thermal conductivity of 138.17 W/m.K and a density of 2.68 g/cm .[3]

The heat collector has dimensions of 580 mm x 1000 mm, with an area of 0.58m^2 , this area will be in contact with the back of the photovoltaic module. The thickness of the aluminium sheet in contact with the photovoltaic module is 1.2 mm.

The heat exchanger model built basically consists of two overlapping aluminium sheets, one smooth and the other crimped. The construction and assembly of the components can be seen in the illustrative drawing in Figure 12.

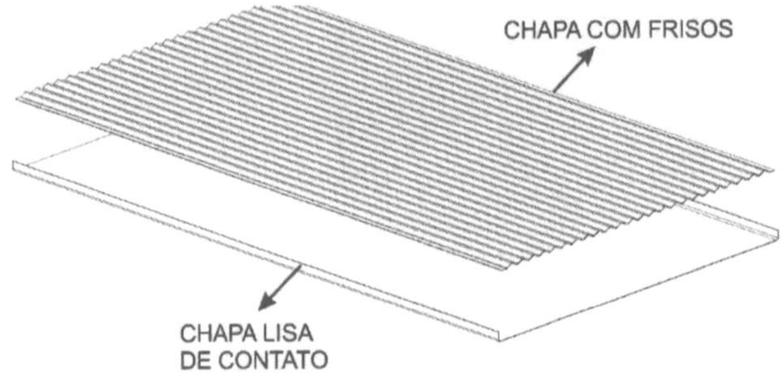

Figure 12Basic form of heat collector construction and assembly

Source: the authors (2016).

Rectangular tubes, also made of aluminium, were used to direct the water through the collector and determine the fluid's outlet and inlet. They were also used to close the collector, as shown in Figure 13. The rectangular tubes were joined to the collector plates using oxyacetylene welding and aluminium rods.

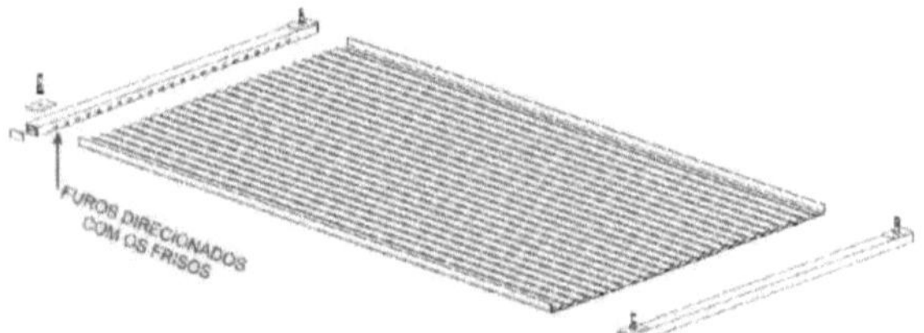

Figure 13 Assembly of the rectangular tubes.

Source: the authors (2016).

The end result was a sturdy aluminium collector, with grooves for draining the water and directing this circulating fluid through the inlets and outlets. Sika polyurethane sealant was used to completely seal the leaks. Figure 14 shows how the heat collector looks: the water supply is at the bottom and the water outlet is at the top.

Figure 14 Heat collectors being tested.
Source: The authors (2016).

To attach the heat collector to the back of the photovoltaic module, carbon steel profiles were used to maintain a rigid and stable structure to support the aluminium collector. Figure 15 shows the shape of the support after assembly.

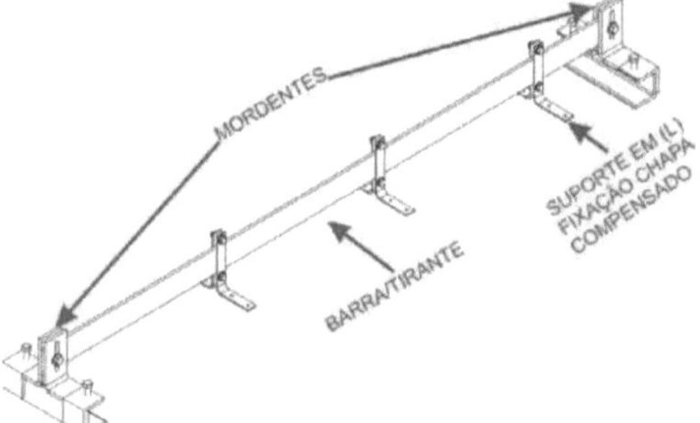

Figure 15 Support model for the aluminium collectors.

Source: The authors (2016)

Two supports similar to those shown in the figure and a %" thick (12.7 mm) plywood sheet with dimensions of 570 mm x 980 mm were used. The final assembly of the collector, plywood sheet and supports can be seen in Figure 16. Glass wool was also used between the heat collector and the wooden sheet, in order to keep the heat collector thermally insulated at the back, leaving only the front face in contact with the photovoltaic module, the area to receive heat transfer.

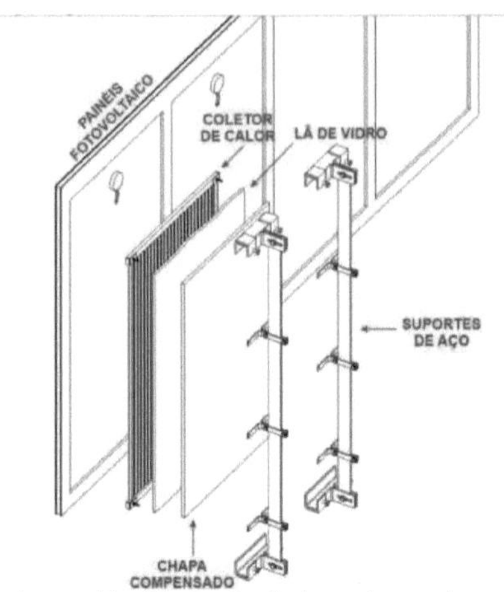

Figure 16 General assembly of the heat collector system and accessories.
Source: The authors (2016).

Figure 17 shows the assembly of the components, with one collector assembled in its working position and the other in the process of being finalised. A transparent 1/2" hose with a wall thickness of 2.0 mm was used. It connects the reservoir to the fluid circulation pump.

Figure 17 Assembly of the heat collectors and accessories.
Source: The authors (2016).

After this assembly, the second collector shown in Figure 17 was deactivated. One of the reasons was the discovery of electrical problems in the module

photovoltaic. This problem was realised after assembly when checking the electrical variables shown in chapter 3.1.7.1 of this work.

3.1.3 Circulation pump

To improve cooling efficiency, a water pump was used to circulate the fluid through the heat collector back to the reservoir. The pump's start-up cycle was programmed using an analogue timer with a start-up cycle every 15 minutes. The pump is a Cubos model Jato 650, with a maximum flow rate of 650 L/h and 120 W power. Its operation is external to the water tank, i.e. it is not submerged. It is widely used for water movement applications in the filtration of ornamental ponds, fountains and artificial waterfalls, fountains or similar. Figure 18 shows where the pump was installed underneath photovoltaic module 01, fixed to the plywood profile.

Figure 18 Circulation pump assembly
Source: The authors (2016).

The maximum flow rate of the pump is around 650 L/h or 10.833 L/min and a maximum pressure of 1.2 mCA. However, as there are losses due to the fluid flowing through the heat collectors and pipes, the pump had to be switched on and its flow rate monitored in accordance with the procedure for filling a standard tank.

Using a graduated reservoir with a maximum capacity of 2 litres, it was timed to fill up completely, with the fluid flowing normally through the collector and the pipe. The values are shown in the table below. Three timed filling tests were carried out and an average flow rate of 4.28 L/min was obtained.

Table 2 Measurements taken to determine the flow rate of the pump used in the experiment.

Measurements	Time (s)		Flow (L/min)
01	29,79		4,03
02	26,92		4,45
03	27,51		4,36
		Average	4,28

3.1.4. Glass wool

Thermal insulators are made up of materials with low conductivity in order to reduce heat transfer. Glass wool was used to insulate the thermal collector from the environment, with a thermal conductivity of 0.038 W/mK. This reduces heat loss through convection, so that the amount of energy (heat) transferred from the module to the water has the lowest possible loss to the environment.

3.1.5. Reservoir

A solar water heating system consists of a reservoir to store the heated water. In this case, the tank was only used to recirculate the water inside the thermal collector.

In order to cool down the photovoltaic module, the water in the reservoir was refreshed every collection day, as it often remained heated from one day to the next, reducing the heat exchange capacity between the module and the water. This procedure also simulated the daily use of the heated water for various activities.

A Styrofoam thermal insulator was used around the tank so that the outside environment would not interfere with the heating or cooling of the water. A rectangular plastic box with a capacity of 150 litres was used. For the assembly process, two 25 mm holes had to be drilled, the upper hole being designated as the return for the heated water coming from the heat collector. The second hole, at the bottom, was designated as the outlet for the water from the reservoir, which would be driven back into the collector by the pump. As shown in Figure 19 below.

Figure 19 Reservoir showing water inlet and outlet.
Source: the authors (2016).

3.1.6 Analogue timer

In order to establish a standardised way of switching the water pump on and off, a 10A G20 127/220V analogue timer was installed in the project, programmed to activate the pump every 15 minutes from 12:00 to 18:00. When the pump is switched off, the temperature of the water inside the heat collector gradually rises. As soon as the timer triggers the circulation, the cold water leaving the reservoir enters the collector, pushing the already heated water back into the reservoir. From 6pm to 12am the following day, the timer did not switch the pump on. The use of this device was extremely important in the experiment, avoiding human error when switching the pump on or off.

3.1.7 Multimeters

To obtain the electrical characteristics of the modules being compared, Impac model IP336-RS digital multimeters were placed on a bench

experimental. Digital multimeters were also used to acquire the current, but they were Minipa model ET 1002.

The electrical data measured was voltage and current, which was saved on a computer with the date and time stamped for each defined time interval. The aim was to graph the behaviour of the photovoltaic modules with the fixed load installed and to determine the point of maximum power. To acquire the voltage data, the test leads were connected in parallel to the load, and to measure the current, the multimeters were installed in series.

The multimeter's main function is to measure the electrical quantities contained in Ohm's law, which are electrical voltage in Vcc or V_{ca} , electrical current intensity in Icc and Ica and electrical resistance in ohm.

The Impac IP336-RS multimeters worked by collecting voltage data in the measurement range of 0 to 60 Vdc. These devices were connected directly to the computer via the RS-232 serial port, with the function of recording the data in a 60 s time interval. The minipa ET 1002 multimeters were set up to collect direct current data in a measurement range of 0 to 10 A. This model has no computer connection, so a web cam was used to take a photo every 60 s and store it on the computer. The accuracies of the devices described can be seen in the annexes. Figure 20 shows the instrument bench installed in the energy laboratory.

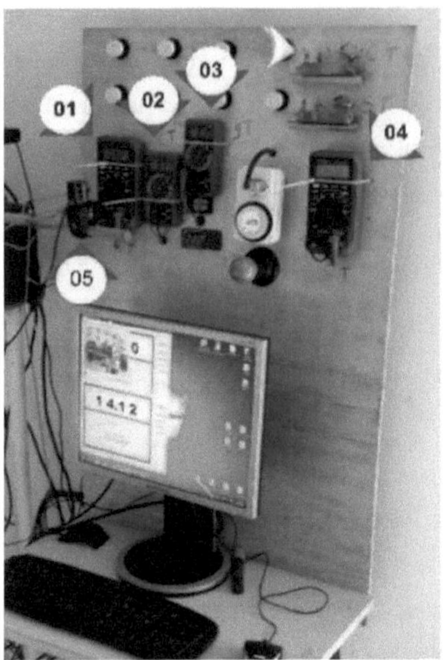

Figure 20 Arrangement of the devices on the bench.

- 01 - Multimeter Impac IP336-RS - Panel voltage P2.
- 02 - Minipa Multimeter ET 1002 - Panel current P1.
- 03-MultimeterMinipa ET 1002- Panel current P2.
- 04-MultimeterImpac IP336-RS- Panel voltage P1.
- 05-WebCamounted for Minipa multimeters.

Source: the authors (2016).

3.1.7.1. Electrical comparison of modules

On 05-03-2016, electrical measurements were taken on the photovoltaic panels in order to determine the electrical characteristics of the installed system. An isolated measurement was made on each photovoltaic module in the renewable energy laboratory. One of the aims of this isolated measurement on each module was to compare the values, to find out the electrical differences that each one has. Firstly, the measurement was taken with no load, i.e. open circuit voltage. The instrument used was an IP 336RS digital multimeter. The modules were named for better understanding and standardisation of the data obtained. The panel with the thermal collector was named P1 and the panel without the thermal collector P2. Another important piece of information is that the heat collector installed in panel 1 was without water, so as not to interfere with the comparison. The panels that are not marked in Figure 21 were not used in the experiment because they had electrical problems due to their physical integrity, i.e. they were malfunctioning.

Figure 21 Numbering of the panels used in the experiment
Source: The Authors, (2016)

To carry out the initial tests, we used a wire resistor from the Telewatt company with 152 ohms and 50 W of power. The wire resistor basically consists of a ceramic tube, which serves as a support for winding a certain length of wire, made of a special alloy to obtain the desired resistance value. The terminals of this wire are connected to clamps attached to the tube. In addition to this, there are other types of construction, as shown in Figure 22.

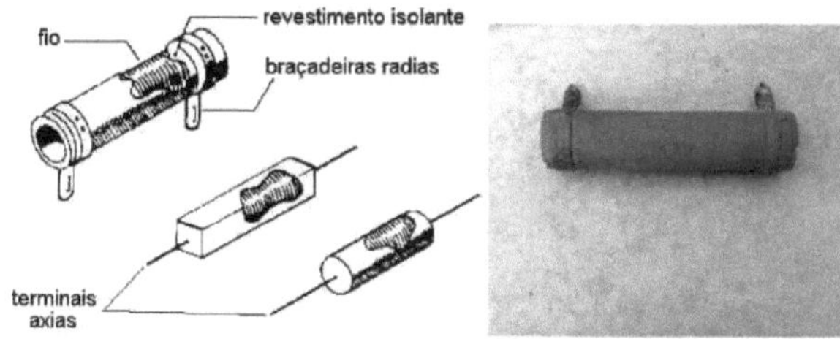

Figure 22 Ceramic wire resistor.
Source: Rocha, 2012.

Before installing the multimeters on the bench, measurements were taken to compare the voltage and current data between the two modules P1 and P2. The results are shown below. The author Rocha (2012) states in his work that Ohm's first law: "The voltage applied across an ohmic circuit is equal to the product of the current and the resistance". From this statement, he is able to apply three more important derived equations used to calculate any of the three electrical quantities voltage, current and resistance with the value of just two parameters. This law is represented by Equation 2:

$$V = R * I \qquad Eq.2$$

where,

V = voltage (d.d.p.) in Volt (V)

I = intensity of electric current in Ampère (A)

R = electrical resistance in Ohm (Q)

- **Measurement on the first module P1:** At the time of measurement there was solar radiation directly incident on the photovoltaic modules of approximately 665 W/m^2 . The measurement started at 10:29 a.m. on 05/03. The module 1 measurement resulted in a value of Vi = 48.1 Vdc and h = 0.316 A. Applying 1ª ohm's law:

$$I = \frac{48{,}1}{152} = 0{,}316\, A$$

- **Measurement in the second module P2:** In the second module the values recorded at 10:37 h were V2 = 48.5 Vdc and I2 = 0.318 A. Applying 1ª ohm's law.

$$I = \frac{48{,}5}{150} = 0{,}319\, A$$

Without the resistive load, the measured voltage was 55.1 Vdc and the current was zero. After connecting the resistor, the voltage at P1 became 48.1 Vdc and the current 0.316 A.

The solar radiation at the time of the P1 measurement was 678 W/m^2 and at the time of the P2 measurement approximately 702 W/m^2 . Figure 23 shows the time of measurement, with the resistive load, of its current for module P1.

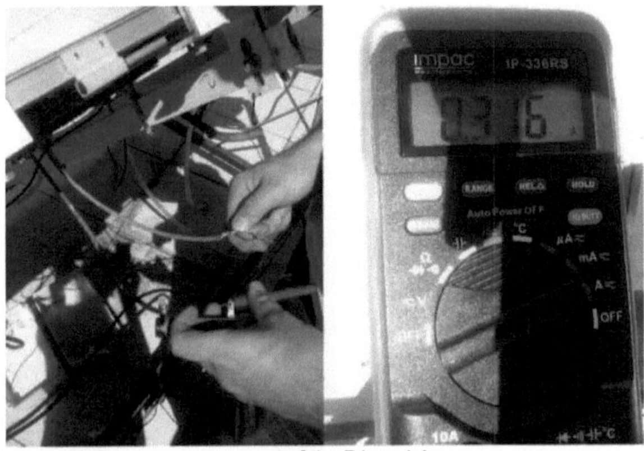

Figure 23 Current measurement of the P1 module.
Source: the authors (2016).

It was therefore found that there was no significant difference in voltage and current between the P1 and P2 modules when the tests were carried out. However, it was found that the other PV modules mounted on the panel showed different values, which is why they were not used.

3.1.8. Digital thermometer

In order to collect the temperatures of the photovoltaic modules and the water circulating through the collector at the inlet and outlet, an Instrutemp 4-channel digital thermometer was used, with the option of recording the data collected in its internal memory or directly on a computer.

The thermometer works with thermocouples of all kinds, but K-type thermocouple sensors were used in the experiment. The sensor's first input channel, called T1, was installed on the front surface of the P2 module. The T2 thermocouple connected to the second channel was glued to the front surface of the P1 photovoltaic module, where the heat collector was inserted. The third thermometer channel was occupied by thermocouple T3 to measure the water inlet temperatures in the collector. Finally, the T4 thermocouple, installed in the last channel of the thermometer, measured the water outlet temperatures of the heat collector.

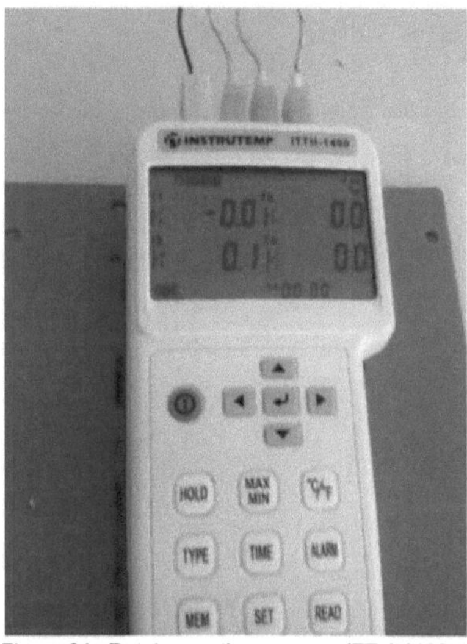

Figura 24 Datalogger thermometer ITTH-1400, Instrutemp brand.
Source: The authors

24.1.8.1. Calibration of the thermometer used

The four thermocouple sensors installed in the project are type K. The ITTH 1400 digital thermometer was calibrated following one of the best known and most effective methods for this procedure.

A container with ice and water at 0° was used to calibrate the digital thermometer. With a bulb thermometer already calibrated according to INMETRO and within its expiry date, the water temperature was visually analysed and found to be 0.1 °C. The sensors were connected to the ITTH digital thermometer and labelled T1, T2, T3 and T4 as shown in Figure 25.

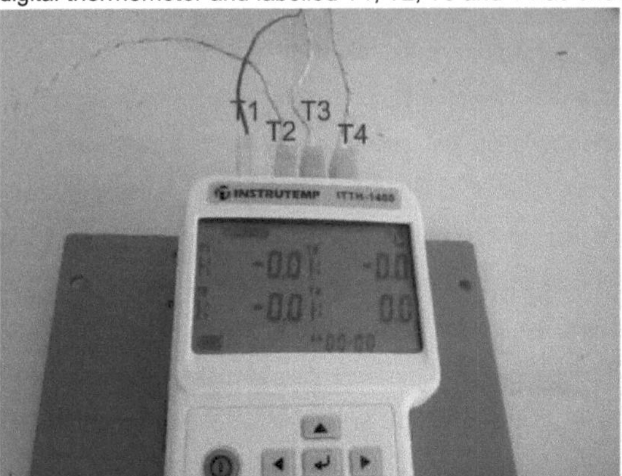

Figura 25 Arrangement of the temperature sensors in the ITTH digital thermometer
Source: The Authors (2016)

The sensors were then immersed in the tub containing the bath at the known temperature. In order to avoid erroneous measurements due to temperature unevenness at different points in the bath, all the sensors were placed inside a tube and immersed in the water for 5 minutes so that all the bodies inserted in the cold water reached thermal equilibrium. The temperatures indicated on the digital thermometer were varying in relation to each other.

The temperatures that were recorded at this time were:

Ti = 2.7 °C;

T_2 = 1.9 °C;

T_3 =0.3 °C;

T_4 =2.3 °C.

Because of this variation in thermocouple measurements, the ITTH 1400 device comes with a factory option to adjust the sensor temperatures by up to 12.7 °C more or less. The individual temperatures of each sensor were then corrected as necessary, so that they were all calibrated to a temperature of 0°C, according to the bulb thermometer. Figure 26 shows the measurement before and after the values were corrected.

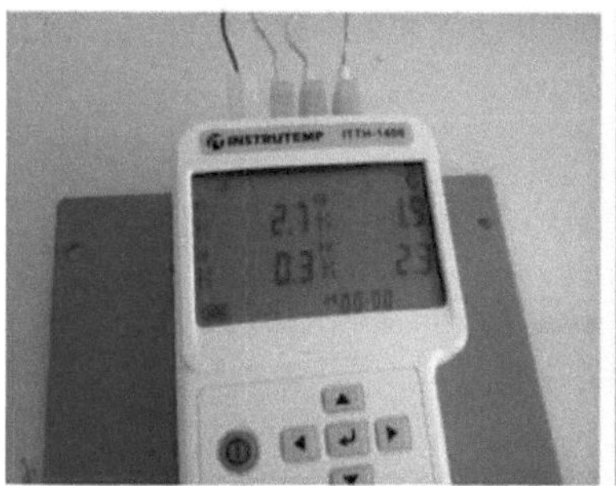

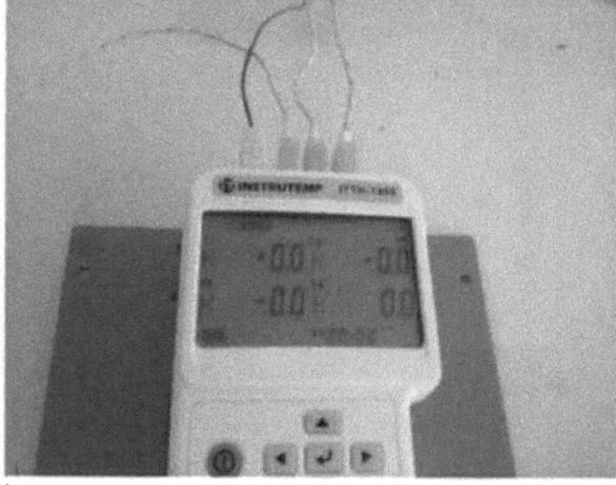

Figure 26 Before and after calibrating the digital thermometer
Source: the authors (2016).

1.2. SOFTWARES

When using the instruments, it is also necessary to use software to manipulate and record the data collected by the instruments. All this software was installed on the computer used, located in the energy laboratory along with all the instruments and the experiment.

1.2.1. Multimeter Datalogger IP336-RS

As the multimeters used to measure the voltages involved in the system have an RS-232 interface for connection to a computer, specific software was used to read the data from these instruments. Figure 27 below shows the interface of the IP336-RS Datalogger programme, supplied by the manufacturer. To start recording the data, the measurement time interval was set at 60 s and the total measurement time was 6 hours per day, from 12 noon to 6 pm. To help read the data, a graph and a list of the measurement data can be displayed.

As two devices were used, each connected to a PV module, it was necessary to run two Datalogger softwares and specify different folders for each programme to store its data. Each multimeter was connected to individual ports on the computer.

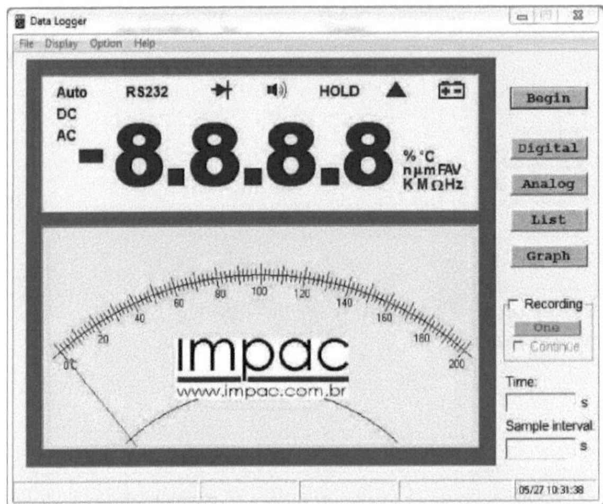

Figure 27 Image of the Impac IP336-RS mutimeter datalogger software
Source: the authors (2016).

To collect the current data, two other multimeters were used, these without an RS-232 or similar connection to save the data directly on the computer. A web camera was installed connected to the computer and directed at the Minipa ET 1002 multimeters, which were side by side. With the aid of the WebCam Xp free software, photos of the devices were programmed to be captured every 60 seconds and stored in the computer's memory, thus not losing the quality of the data collected, as the collection interval and time were the same as for the IP336-RS multimeters.

Figure 28 Photo taken on 29-03-2016 using the webcam Xp free software.
Source: the authors (2016).

3.2.2. Temperature datalogger software

The ITTH 1400 digital thermometer used in the experiment has data acquisition software called Run D4IThermoMeter, where you can save the temperatures measured in the programmed time interval directly to your computer. It has an easy-to-use and informative interface which shows the temperatures updated in real time from the reading. With this software it is also possible to store the maximum or minimum temperature peaks between the time intervals programmed in the save. The interval time between recording temperatures was fixed at 60 s, so all the devices and software are programmed for the same intervals, times and dates when the values are acquired.

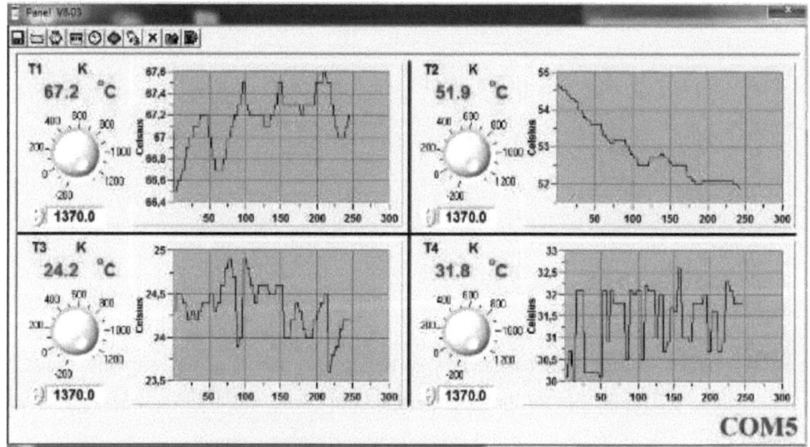

Figure 29 Screenshot of the Run D4IThermoMeter software
Source: Datalogger and RS232 Interface.

Figure 29 shows the interface of the temperature data acquisition software. There are 4 windows, one for each thermocouple sensor channel and the model of thermocouple used, in this case all type K, is also displayed, as well as the temperature being measured at that moment. The same image also instantly generates a graph of the temperature profile over time. Thanks to this software and the 4-channel digital thermometer, it was possible to reliably acquire the temperature data of the system under study.

3.3. POWER CIRCUIT FOR THE EXPERIMENT

In order to analyse the electrical characteristics of the photovoltaic panels, devices capable of consuming the power produced by the modules were assembled, so that electrical quantities such as voltage and current could be measured.

The electrical circuit consists of a 60 ohm, 30 W wire resistor and two 12 V light bulbs each with a resistance of 340 ohms. The same resistive load devices were used for the two panels in order to compare the electrical characteristics of both. The diagram shown in Figure 30 shows how the electrical circuit described looks.

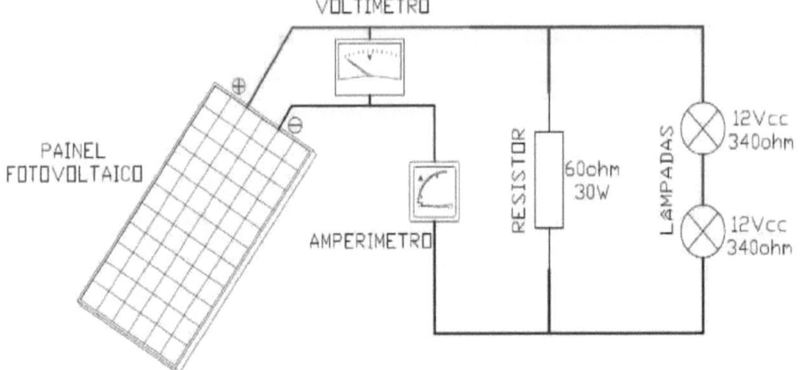

Figure 30 Representative diagram of the electrical circuit
Source: The authors (2016).

The wiring diagram shows the connection of the resistors and bulbs, with the resistor connected in parallel to the panel and the bulbs in series. In this way, the equivalent resistance of the circuit was 55 ohms. Figure 31 shows the instrument bench after assembly.

At the top is the power circuit with resistor and bulbs for the photovoltaic module P1 with the thermal collector installed. Below is the electrical circuit for the module without the thermal collector P2. The multimeters that read the module with the thermal collector are on the right of Figure 31. The multimeters that measure the electrical characteristics of the panel without a thermal collector are on the left of the figure.

This panel also has a blue light to indicate that the water pump is on. The appendices to this paper contain a general schematic drawing of the experiment bench. This drawing shows separately the two sets of electrical diagrams for each PV module, as well as the layout of the instruments and sensors used.

Figure 31 shows the following components:

- 01 and 02 - Ceramic wire resistor - 60 ohms, 30 W;
- 03 - 12 V bulbs in series - Circuit P1;
- 04 - 12 V bulbs in series - Circuit P2;
- 05 - Analogue timer - Switches water pump on/off;
- 06 - Indicator light when the pump is switched on.

Figure 31 Experiment bench

Source: The authors (2016).

1.4. ORIENTATION AND INCLINATION OF THE PHOTOVOLTAIC PANEL

The bibliographical references showed that for proper and efficient operation, the photovoltaic panels should be orientated in the direction of geographic north. As the installation site of the panels was interfered with by shading from the building, block H of Unoesc Xanxerê, in the morning and in the afternoon from the trees located next to the Unoesc block, they were positioned to obtain the greatest possible solar incidence.

The angle of inclination of the photovoltaic module platform was 45° in relation to the ground.

This is not the most suitable inclination, but it was decided to maintain this inclination because the panel support in the renewable energy laboratory was like this.

Figure 32 Installation of the modules showing their routing.

Source: The Authors (2016)

Figure 32 shows the line of shading caused by the parapet. A small area of vegetation can also be seen in the background, which after 4pm interferes with the intense shading of the modules.

1.5. EFFICIENCY CALCULATIONS

Photovoltaic cells cannot convert all the incident solar radiation into electricity, due to the limitations of the technology and the losses inherent in the process. The efficiency of a photovoltaic module is defined as the ratio between the maximum electrical power generated and the power incident on it.

The following equations were used to determine solar conversion efficiency, according to the author Marques, (2008):

$$\eta = \frac{Pger}{Rinc} \qquad Eq. 3$$

Where:

η: Electrical Efficiency

Pger: Electric Power Generated (W)

Rinc: Incident Radiation panel (W)

$$P_{ger} = Is \; x \; Vs \qquad Eq. 4$$

Pger: Electric Power Generated (W)

Is = Measured electric current (A)

V_s = Measured electrical voltage (V)

$$R_{inc} = G \, x \, A \qquad Eq. 5$$

Rinc: Incident Radiation panel (W)

G = Solar radiation (W/m)^2

A = Panel area (m).^{2}

CHAPTER 4

RESULTS AND DISCUSSIONS

The system was monitored from 29/03/2016 until 19/05/2016, although not all days were used due to weather variations, totalling 24 days of data collection. The measurements were taken from 12:00 pm to 6:00 pm, with data recorded every 60 seconds. This generated a total of 360 daily data recordings. Over the 24-day collection period, a total of 8,640 voltage, current, temperature and radiation data points were recorded. Table 3 below shows the summarised results. C.T (Panel with thermal collector) and S.T (Without thermal collector).

Table 3 General summary showing the daily averages of the data collected.

	CURRENT (A)		VOLTAGE (V)		ENERGY GENERATED (W)		EFFICIENCY %		DIFFERENCE BETWEEN ENERGY GENERATED %	MEDIUM RADIATION W/m²
DATE	**C.T**	S.T	C.T	S.T	**C.T**	**S.T**	**C.T**	S.T		
29/03/2016	0,72	0,7	43,43	41,83	31,44	29,63	4,90%	4,55%	**5,78%**	804,56
30/03/2016	0,6	0,6	31,02	27,31	20,35	17,82	3,18%	2,85%	**12,45%**	782,76
31/03/2016	0,32	0,3	19,08	18,48	10,87	10,59	2,14%	2,03%	**2,62%**	488,96
01/04/2016	0,71	0,72	42,7	41,22	30,71	29,91	5,27%	5,14%	**2,60%**	766,28
04/04/2016	0,59	0,61	36,18	34,67	24,91	24,61	3,73%	3,78%	**1,22%**	754,2
05/04/2016	0,66	0,63	38,68	36,98	26,06	24,17	4,41%	4,16%	**7,25%**	716,96
06/04/2016	0,7	0,69	42,01	40,56	30,27	28,97	5,27%	4,99%	**4,29%**	724,84
13/04/2016	0,45	0,44	27,37	25,16	15,44	14,54	4,43%	3,91%	**5,80%**	548,92
14/04/2016	0,3	0,27	18,38	16,23	**7,17**	5,46	2,23%	1,78%	**23,94%**	381,4
15/04/2016	0,65	0,63	38,68	36,78	27,74	26,06	4,91%	4,43%	**6,07%**	704,08
18/04/2016	0,63	0,63	37,73	35,67	25,83	24,77	5,28%	4,76%	**4,13%**	600,21
19/04/2016	0,51	0,49	30,9	28,92	20,25	18,92	3,02%	2,65%	**6,58%**	696,2
20/04/2016	0,61	0,59	36,5	34,73	25,9	24,48	4,73%	4,61%	**5,49%**	613,36
23/04/2016	0,68	0,67	40,96	39,27	30,03	28,62	5,63%	5,53%	**4,69%**	621,8
29/04/2016	0,6	0,59	36,12	35,29	25,72	24,39	5,25%	4,98%	**5,20%**	613,94
02/05/2016	0,65	0,63	38,28	37,38	27,88	26,82	5,91%	5,57%	**3,79%**	592,56
04/05/2016	0,65	0,63	39,19	37,25	28,29	26,47	5,89%	5,31%	**6,45%**	555
05/05/2016	0,6	0,58	36,21	34,28	25,41	23,81	5,13%	4,52%	**6,31%**	543,68
07/05/2016	0,34	0,31	20,59	18,94	10,6	9,53	2,49%	2,16%	**10,09%**	387,21
10/05/2016	0,41	0,38	25,1	23,05	14,32	12,86	3,03%	2,58%	**10,21%**	452,84
14/05/2016	0,59	0,58	35,53	34,03	25,06	23,88	4,54%	4,26%	**4,70%**	603,46
17/05/2016	0,66	0,64	39,74	37,87	28,96	27,12	6,00%	5,63%	**6,34%**	537,4
18/05/2016	0,66	0,64	39,42	37,64	28,51	26,91	5,97%	5,67%	**5,61%**	530,52
19/05/2016	0,64	0,63	38,56	36,93	27,58	26,21	5,78%	5,49%	**4,97%**	541,96
AVERAGE	0,58	0,57	34,6	32,94	23,72	22,35	4,55%	4,22%	**6,52%**	606,8
				SOMA	**569,3**	**536,5**				

Source: The Authors, (2016).

The difference in energy generated between modules P1 and P2 varied by an average of 6.52 per cent. The efficiency of the panel with the heat collector averaged 4.55%, while the efficiency of the solar panel without the collector averaged 4.22%. This variation suggests that the P1 CT module produces more energy when it is cooled.

14/04/2016 was the day on which the least energy was accounted for due to the heavy cloud cover. The average energy generation in the P1 CT module was 7.17 W, while in P2 ST the generation was 5.46 W. Even on adverse days without much radiation, it was possible to generate more energy in the module with the heat collector.

Table 3 shows a general summary of all the days on which data was collected and all the variables analysed and presented as averages. On 04/04/16 the energy conversion efficiencies were higher in P2 ST. As the differences are small, we can conclude that there was some interference in the data collection due to uncontrolled factors.

The P1 CT module generated the most energy during the collection days, totalling 569.3 W, while the P2 ST module generated 536.5 W, the difference in energy generated being 32.8 W.

The experiment showed that as the temperature of the photovoltaic module varies, its voltage changes. In the photovoltaic module in which the heat collector was installed, it can be seen that its voltage was higher than the voltage of the conventional panel, both of which were affected by the same solar radiation. Figure 35 shows an image generated by the data collection software on 04/05/2016.

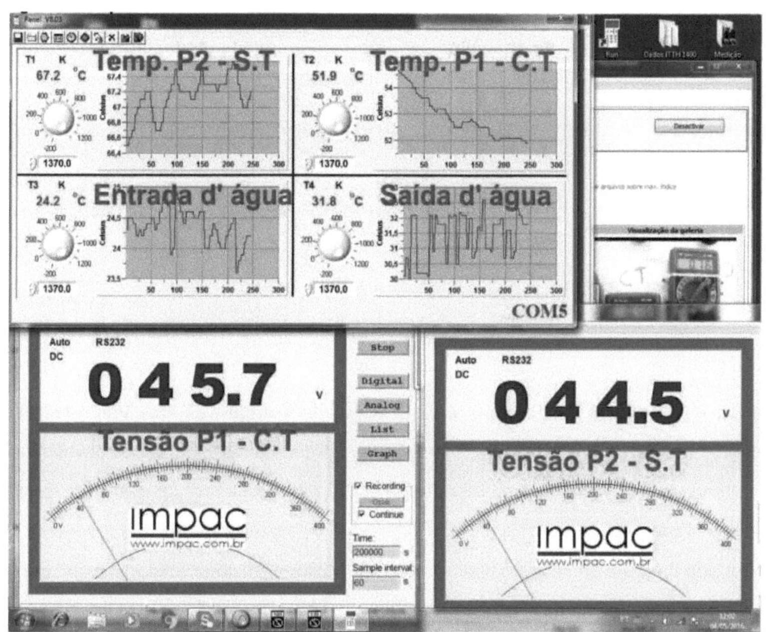

Figure 33 Screenshot from 04/05/2016

Source: The authors (2016).

In Figure 33, the P2 ST module had a temperature of 67.2 °C, while P1 CT had a temperature of 51.9 °C, a difference of 15.3 °C. Their generated voltages were measured as shown in the figure at P1 with 45.7 Vdc and P2 with 44.5 Vdc, a difference of 2.62 % between the voltages, so the P1 panel, which had a temperature approximately 15°C lower than P2, generated 1.2 V_{cc} more than P2 ST, a decrease of around 0.2%/°C.

When the modules are at the same operating temperature, their voltage is equalised. This shows the influence that temperature has on the solar module's operating voltage. There is also a slight variation in the current strength, but it is less intense than the voltage.

Figure 34 shows the voltage and temperature measurements with the circulation pump switched off, i.e. without any type of cooling, leaving photovoltaic modules P1 and P2 at the same temperature.

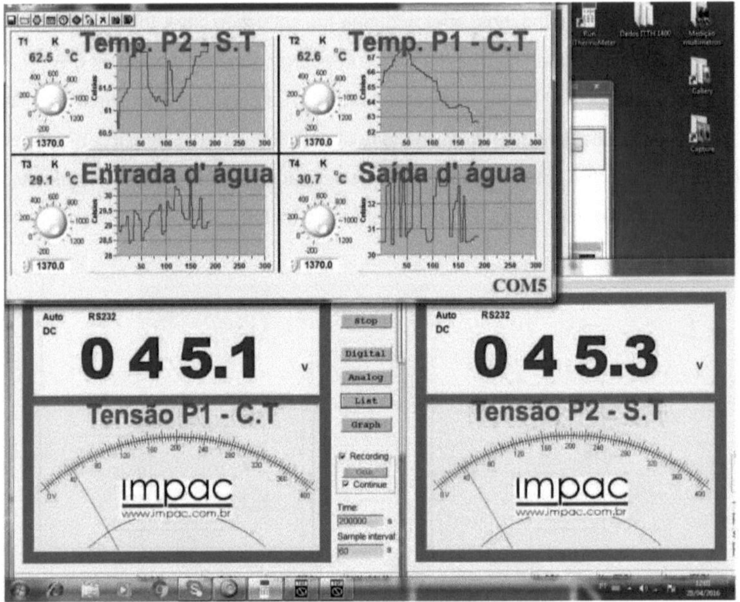

Figure 34 Screenshot of 20/04/16.
Source: The Authors, (2016)

In Figure 34, P1 CT had a temperature of 62.6 °C and its voltage was 45.1 Vdc, while on panel P2 its temperature was 62.5 °C and its voltage was 45.3 Vdc, a small difference between the modules, this difference being within the precision range of the multimeters, or caused by external interference that was not measured.

After processing the data and putting it into Excel spreadsheets, clearer interpretations were achieved in terms of the generation efficiency in both cases, and the construction of graphs to better demonstrate the results. As can be seen in Graph 1, on 29/03/2016, the first day of data collection, an almost perfect curve of the voltage varying according to solar radiation was generated. The sky was cloud-free on this day. As can be seen in the graph, the difference between the voltages generated by the modules is remarkable.

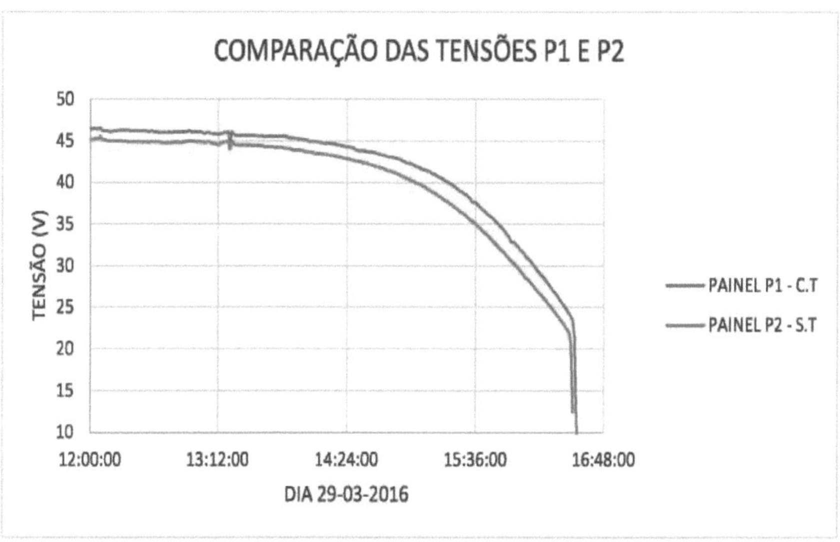

Graph 1 Comparison of the voltages generated between P1 CT and P2 ST - Day 29/03/2016.

On this day, the average voltage for P1 CT was 43.43 Vdc and the average for P2 ST was 41.83 Vdc, a percentage difference between the voltages of 3.68 per cent. As already mentioned, the measurements were taken from midday until 6pm, and after 4pm the shading at the module installation site was very intense due to the trees next to the laboratory.

Therefore, in order to calculate the averages, we used the data from 12:00 pm to 4:30 pm. This shading can be seen in the graph, which shows the sharp drop in the curve after 4pm. On the same day, the average radiation was 804.56 W/m^2 and the maximum radiation peak was 963 W/m^2 at 12:10 pm.

29/03/2016 was the day with the highest energy generation, due to the weather conditions and the high solar incidence. When the sun's rays decrease, causing the panel to be shaded, the result shown in Graph 2 is obtained.

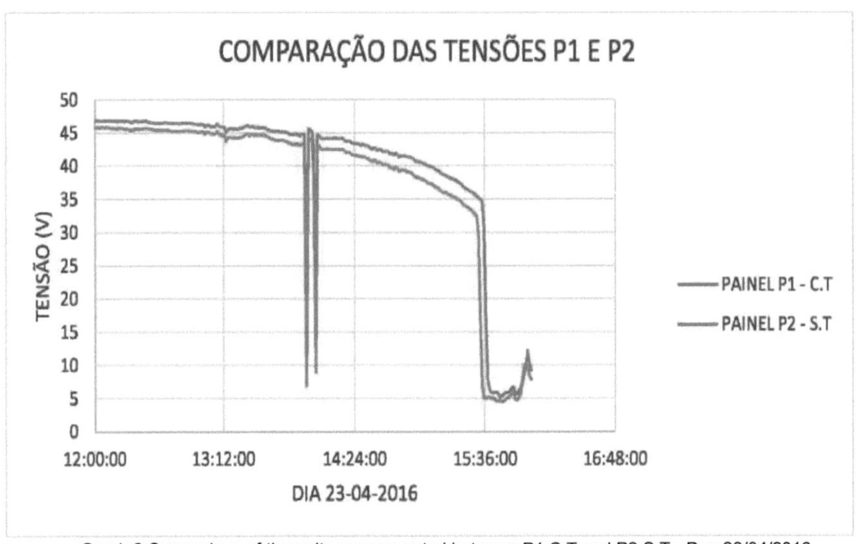

Graph 2 Comparison of the voltages generated between P1 C.T and P2 S.T - Day 23/04/2016.

At around 2pm on 23/04/2016 there was a very sharp drop in the voltage generated, due to the sudden decrease in radiation at that time. This is usually due to the passage of very thick clouds. After this event, voltage generation began to rise again and the generation curve continued.

4.1 CURRENT-VOLTAGE CURVE

There is a huge variety of computer tools for analysing or designing photovoltaic systems. To draw the characteristic curve of the amorphous silicon photovoltaic panel, we used the Crearray V1.0 software developed by LABSOL (Solar Energy Laboratory) located on the campus of the Federal University of Rio Grande do Sul, which tests equipment and develops software.

Figure 35 shows the I-V curves of photovoltaic modules P1 and P2. The data in Table 1 was used to indicate the module's electrical characteristics. The maximum radiation recorded on 29/03 was 963 W/m^2 and the temperature of the P2 ST module was 69.7 °C at 12:08 pm and 56 °C for the P1 CT module.

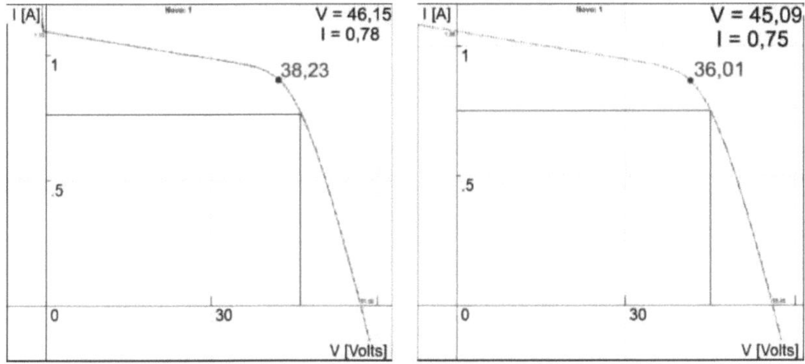

Figure 35 Characteristic curves of the P1 and P2 modules.
Source: The Authors, (2016).

The graph on the left shows the maximum power point at 38.23 W for the P1 CT module. It can be seen that at that moment the meeting point of the 46.15 V voltage line and 0.78 A current was being used, giving a power consumption by the installed load of approximately 36 W, slightly below the maximum power point. For the graph on the right, referring to the P2 ST module, the maximum power point was around 36.01 W and the power used at that time was 45.09 V voltage and 0.75 A current, resulting in a power of 33.8 W. The radiation used in the Crearray software for both graphs was 963 W/m^2 . The difference lies in the temperature of the modules, which is why the curves showed different curves and points of maximum power.

4.2. GENERATED POWER AND RADIATION

Graph 3 below shows the relationship between the power generated and the variation in the incidence of radiation on the solar panel. As expected, the days with the highest radiation also see the highest power generation. It is also possible to see the radiation line decreasing over the days and months due to the passing of the seasons and consequently the decrease in solar radiation. Almost every day, electricity generation in the P2 ST module was lower than in the P1 CT module.

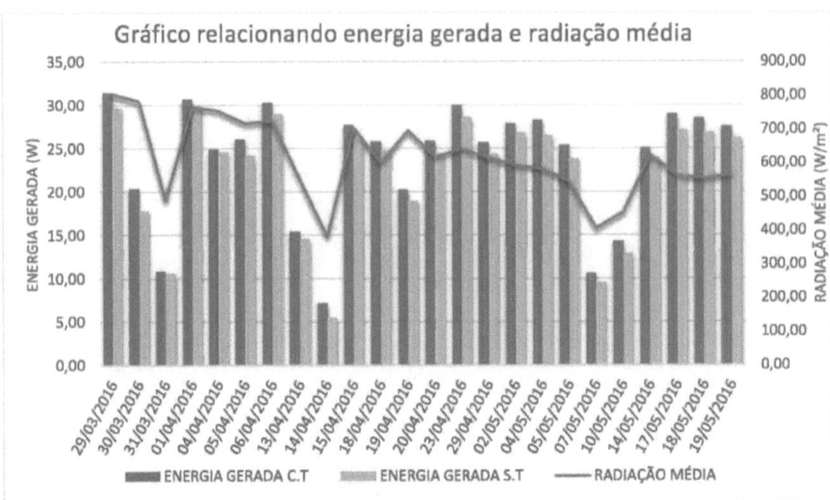

Graph 3 Graph of energy generated x incident radiation.

There is a correlation between the energy generated and the incident radiation. Even on days with little direct radiation, electricity was generated, proving the concept that it can also be generated with diffuse radiation. According to the graph, the days with the highest energy generation were 29/03, 01/04, 06/04 and 23/04. The average generation for these days was 30.61 W for P1 CT and 29.28 W for P2 ST.

The days with the lowest electrical generation were 31/03, 13/04, 14/04, 07/05 and 10/05 respectively. The average generation for these days was 11.68 W for P1 CT and 10.59 W for P2 ST. The overall average power for all the days was 23.72 W for the module with the heat collector installed and 22.35 W for the P2 photovoltaic module which was operating conventionally without cooling.

4.3 HEAT GENERATION

Through this experiment it was possible to realise that the water circulating in the heat collector was heated to temperatures of up to 40 °C on certain days. By using temperature sensors at strategic points, it was possible to extract a quantitative result of the equipment's performance. The author Marques, (2012) mentions in his work that the thermal efficiency of a hybrid panel is lower than that of a solar panel.

than a conventional solar thermal collector, due to the performance losses caused by the fact that the sun's rays do not reach the collector directly. For the radiant energy to reach the heat collector, it must pass through a series of thermal resistors, as shown in Figure 36.

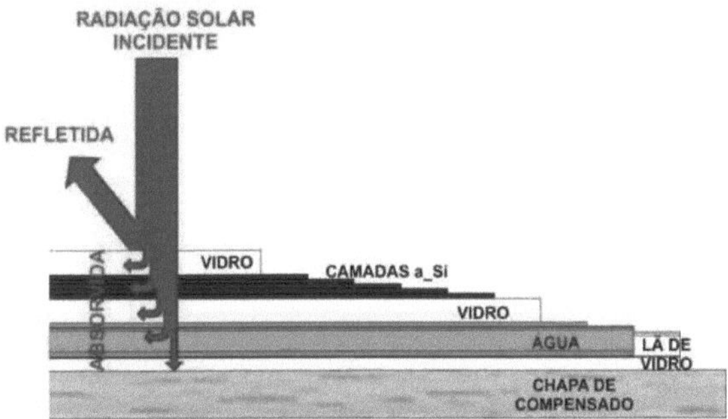

Figure 36 Simple demonstration of radiation distribution

Source: The Authors (2016)

With the data, it was possible to find the average temperatures in the 12:00 to 16:00 range, as was done with the electrical data. The graph for 29/03 shows the temperature profile of modules P1 and P2, as well as the water entering and leaving the heat collector. The temperature line of the P1 module, which was cooled by the heat collector, was below the temperature of the conventional P2 module for almost the entire time period, showing that some of the heat generated was transmitted to the fluid. After 4.30 p.m. the temperatures dropped considerably, and on all days there was an inversion of temperatures, where P2 S.T. had its temperatures marked below P1.

This is because the water circulating through the thermal collector is in a closed system, where at the end of the day the heated water keeps module P1 warmer than P2. You can see from this graph that the water entering the collector was at a lower temperature at the start of the data collection. After a certain period of time, from around 3.30pm to 5pm, there was a balance in the temperatures, because the water inlet temperatures, after heating, are very close to the water outlet temperatures of the collector. In this case, the water was entering at a temperature of approximately 34 °C and leaving at approximately 35 °C. All of the above can be seen in Graph 4.

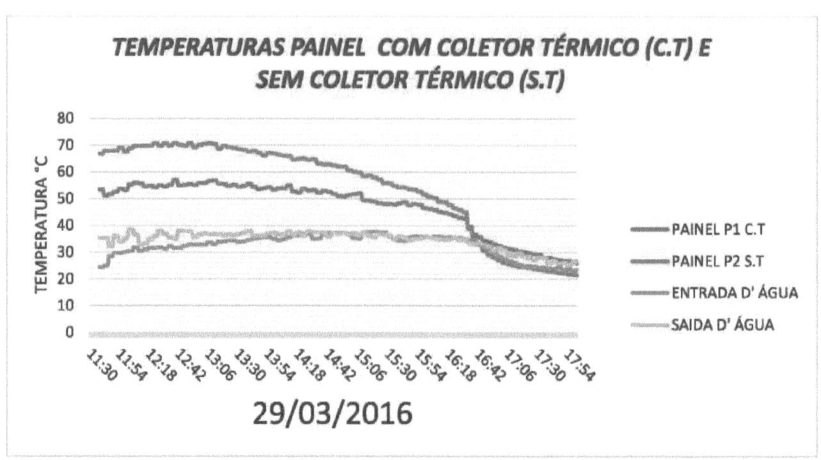

Graph 4 Panel temperatures with thermal collector (C.T) and without thermal collector (S.T)

Therefore, the water outlet temperature remains warm for a period of time, as the water tank was lined with insulating material (Styrofoam), reducing heat loss. On 29 March, the measured temperature data was averaged, giving 36.26 °C at the water outlet of the heat collector. On this day, the water pump was switched on at 12:00 p.m. where the temperature at the inlet to the collector was 24.4 °C and at 4:00 p.m. it was leaving at a temperature of 36.0 °C.

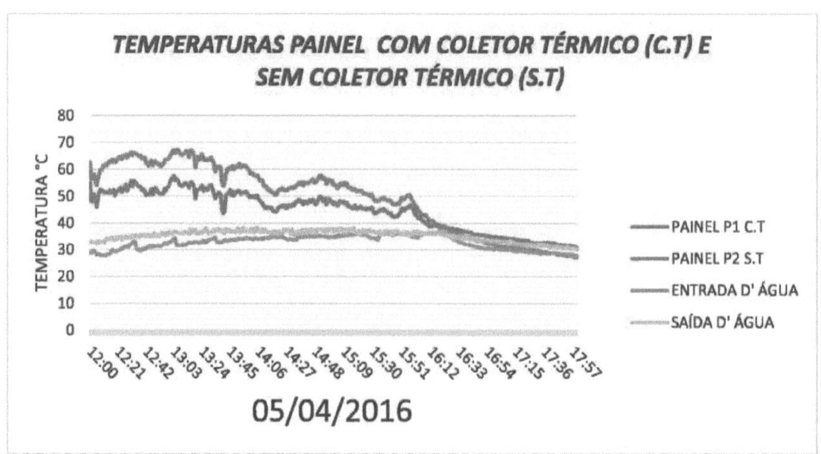

Graph 5 Panel temperatures with thermal collector (c.t) and without thermal collector (s.t)

Graph 5 for 05/04 shows the same condition as the previous graph, where the temperature of the P2 S.T module was higher than the others practically all the time, reversing its position after 16:30 when the radiation is already very low due to interference from the shading that is caused from that time onwards. On this day, the water reached 38.2 °C at two o'clock in the afternoon. From 4.10 p.m. onwards, the temperatures of the water entering and leaving the heat collector became very close, resulting in a thermal equilibrium due to the water in the thermal reservoir being hot and returning to the collector in this way, thus reducing the gradient between the temperatures of the water and the collector and causing this equilibrium. Table 4 below summarises all the days on which temperatures were taken.

Table 4 summarises all the days on which temperatures were taken

	Panel P2 S.T. (T1) "C	Panel P1 T.C. (T2) °C	Water inlet (T3)°C	Water Balance (T4) 'C	Ambient Temp. °C
29/03/2016	64,86	52,83	33,77	36,30	23,61
30/03/2016	53,87	43,46	29,37	33,53	27,38
31/03/2016	45,12	40,95	32,50	33,22	26,30
01/04/2016	53,34	43,82	32,42	35,39	29,22
04/04/2016	57,93	49,95	33,87	37,14	31,06
05/04/2016	57,56	49,20	33,35	36,23	29,67
06/04/2016	60,52	52,59	36,51	36,80	30,64
13/04/2016	44,46	40,08	29,79	31,93	28,36
14/04/2016	37,53	34,71	27,71	28,95	25,33
15/04/2016	54,36	47,73	32,57	34,22	29,75
18/04/2016	61,46	52,57	32,78	35,56	29,61
19/04/2016	47,97	43,29	30,89	32,50	28,98
20/04/2016	55,56	48,43	32,50	35,06	29,72
23/04/2016	56,28	49,52	28,08	30,32	28,15
29/04/2016	47,30	36,58	19,78	22,75	15,10
02/05/2016	54,73	46,26	27,58	30,40	21,70
04/05/2016	58,67	48,67	30,09	34,44	21,81
05/05/2016	56,23	46,50	27,72	33,61	22,99
07/05/2016	37,32	35,42	21,11	21,86	20,69
10/05/2016	41,55	35,12	23,27	26,10	20,11
14/05/2016	46,87	41,25	26,68	27,29	22,13
17/05/2016	50,70	39,86	21,67	25,08	12,26
18/05/2016	52,20	42,88	25,93	27,91	14,58
19/05/2016	49,24	42,90	26,70	26,86	19,27

Source: The Authors, (2016).

According to the table, the highest outlet temperatures from the collector were on 29/03, 04/04, 05/04, 06/04 and 18/04. On these days the average water temperature reached 37°C, within the ideal range for bathing. There is also a relationship between ambient temperature and heat generation. The wind has an influence on the temperatures of the panel and the collector, as its flow over the experiment causes the surface to cool slightly by convection. The overall average of the outlet temperatures of the water heated by the heat collector was 31.39°C. This means that it is not feasible to use a heat exchanger to utilise heated water directly for any purpose. In order to use it, another form of auxiliary heating must be used to achieve a temperature suitable for bathing, for example.

CHAPTER 5

CONCLUSION

The insertion of a thermal collector into a common photovoltaic generation system has brought some advantages to the system. It can be concluded that the cooling of an amorphous silicon photovoltaic panel tends to generate more electricity than a conventional solar panel. This improvement can be seen in the presentation of the voltage graphs, where it was possible to visualise the difference between the voltages of the photovoltaic modules. A large number of electrical and temperature records were acquired to prove the experiment.

It was concluded that the difference in energy generated between modules P1 and P2 varied by an average of 6.52 per cent, showing a trend towards greater generation in the module with the heat collector. The electrical efficiency in relation to the incident radiation was demonstrated for both modules, and the module with the heat exchanger was found to have an average of 7.25 per cent more electrical efficiency than the conventional system.

With this experiment we were also able to obtain heated water. Heat generation was considered low because the temperatures reached by the water were not sufficient for storage during periods without solar radiation and subsequent use. There were days when the temperature was enough for a hot shower, but due to thermal losses at the end of the day it would need to be heated again to be used.

It was concluded that the heat collector has a low heat exchange efficiency due to the construction model not being the most suitable. A suggestion for future work would be to improve the thermal insulation of the pipes and reservoir and also to improve the heat exchange capacity of the heat collector in order to obtain a greater amount of heat. It may also be possible to calculate the thermal efficiency of the system.

CHAPTER 6

REFERENCES

ABRAVA. **Training Manual for Solar Heating System Projects.** April 2008 edition.

ALMEIDA, Laura Manuela Melo Carvalho Esteves; Dissertation - **Study of a hybrid solar collector for the production of electricity and heat.** University of Porto, 2008.

ANDRADE, Airton Cabral de; PhD thesis: **Analysis and simulation of temperature distribution in photovoltaic modules,** UFRGS-RS 2008

ANEEL. **Atlas of Electricity in Brazil.** National Electric Energy Agency, Brasília - DF, Brazil, 2010.

BORTOLETTO, D. R., PESSUTO, C. C.: **Performance analysis of a low-cost solar heater;** Alternative thermal reservoir. PUC Campinas, 2012.

BUHLER, Alexandre José. **Determination of photovoltaic parameters based on characteristic curve tests without lighting.** 2007.

CÂMARA, Carlos Fernando. **Grid-connected photovoltaic systems** Monograph. LAVRAS, MG. 2011.

CARVALHO, Marcelo de Fraga. **Angular Sensor Based on Hall Effect. XX f.** Course Conclusion Work in Electrical Engineering - Department of Electrical Engineering. Lutheran University of Brazil. Canoas, RS. 2011.

CHARALAMBOUS, P. G., KALOGIROU, S.A., MAIDMENT G., **KARAYIANNIST.G.,"Photovoltaic Thermal (PV/T) Collectors: A Review",** HPC 2004 - 3rdInternational Conference on Heat Powered Cycles, Cyprus, October2004;

Datalogger and RS232 Interface. Available at <http://docs-europe.electrocomponents.com/webdocs/1188/0900766b81188a0a.pdf>. Accessed on 15 March 2016.

DUARTE, Pedro Nuno Alves. **Solar Thermal Collectors - Analysing the thermal performance of technological solutions on the market.** Integrated Master's Dissertation in Mechanical Engineering. Faculty of Engineering, University of Porto. July 2012;

Solar Thermal Energy. Available at: <http://www.solaribérico.pt/energias/solar- termico/>. Accessed on: 22 April 16.

FILHO, Geraldo Lúcio Tiago; JÚNIOR, Ângelo Stano. **Renewable Energy Series: Renewable Energies.** Itajubá, 2007.

Supplier of modules for thermal sensors. Available at: <http://www.veder.com.br/catalogo-de-produtos/modulos-para-aquisicao-de-dados- via-rs485/>. Accessed on 25 May 2015.

GUARIENTE, Laércio José. **Comparative study of the performance of a solar water heating system with natural circulation and forced circulation.** Postgraduate dissertation. Londrina-PR, 2005.

GUENA, Ana Maria de Oliveira. **Environmental assessment of different forms of electricity generation.** Master's thesis, IPEN - Nuclear and Energy Research Institute, 2007.

GUIMARÃES, Mário Jorge Cardoso. **Behaviour of Si-a photovoltaic solar cells in the field and their degradation.** 2010. PhD Thesis. Faculty of Science and Technology.

HECKTHEUER, L. A.; **Analysing Photovoltaic Module Associations;** 2001. Thesis (doctorate in mechanical engineering) - Federal University of Rio Grande do Sul; Porto Alegre, 2001. Available at: < http://www.seer.ufv.br/seer/index.php/reveng/article/viewFile/141/67>. Accessed on: 14 September 2015.

INGEMECÁNICA. **Solar Photovoltaic Installation for Housing -** Tutorial no. 192. Available at: <http://ingemecanica.com/tutorialsemanal/tutorialn192.html>. Accessed on: 05 October 2015.

KUGLER, Henrique. **Under the rays of the sun,** 09 Feb.2010. Science Today. Available at: <http://cienciahoje.uol.com.br/noticias/2010/02/sob-os-raios-do-sol>.

Glass wool. Available at: <http://wwwo.metalica.com.br/la-de-vidro-isolamento- temnico-e-acustico>. Accessed on: 03 March 2016.

LOPEZ, Ricardo Aldabó. **Solar energy for electricity production.** Editora Artliber, São Paulo, 2012.

MACHADO, Carolina T.; MIRANDA, Fabio S. **Photovoltaic Solar Energy: A Brief Review.** Revista Virtual de Química, v. 7, n. 1, p. 126-143, 2014.

MARQUES, Ricardo Licínio Moreira da Rocha. **Feasibility assessment of hybrid photovoltaic and thermal collectors for application to water heating and electricity micro-generation.** MIEM Dissertation Report.
Faculty of Engineering, University of Porto. Integrated Master's Degree in Mechanical Engineering. Portugal, 2008.

MATIAS, Vandeir Robson da Silva **Energy and the Environment: Geostatistical Applications.** Minas Gerais Geography Symposium - Alfenas. MG, 2014.

MOREIRA, Lúcia. **Temperature measurement using thermocouples.** Cerâmica Industrial, v. 7, n. 5, p. 51-53, 2002.

NILSSON, J. W.; RIEDEL, S. A. **Electric Circuits.** Ninth edition. [S.l.]: Pearson, 2011.

NUNES, Otacilio José de Macêdo. **Feasibility study for alternative electrification of a residential consumer unit in fortaleza-ce using a photovoltaic system.** State University of Ceará. CE, 2011
PALZ, Wolfgang. **Solar Energy and Alternative Sources.** Publisher, Hermus. São Paulo, 2012.

PINHO, João Tavares; et al. **Engineering manual for photovoltaic systems,** CEPEL - CRESESB, RJ, 2014;

PRIEB, César Wilhelm Massen. **Development of a photovoltaic module testing system.** 2002.

RODRIGUES, Pedro Rabela. **Design of an experimental teaching bench for the use of solar and wind energy.** UnB - Brasília, DF. 2014;

ROGANI, B. C. **Modelling of a Grid-Connected Photovoltaic System and Performance**

Analysis under Shading Conditions and Temperature Variations. University of São Paulo, São Carlos, 2013.

RORIZ, Luis. **Solar Thermal: Use of solar thermal panels.** 2008.

RÚTHER, Ricardo. **Photovoltaic Solar Buildings.** Editora UFSC / LABSOLAR. Florianópolis, 2004.

STANO JÚNIOR, Ângelo; TIAGO FILHO, Geraldo L. **Energias Renováveis.** 1ª Edition, Fapepe, 2007.

TEIXEIRA, Tiago Rafael Contins. **Study of a hybrid system with thermo-photovoltaic solar collectors coupled to an electric thermo-generator.**
Master's dissertation in Mechanical Engineering. University of Porto, 2012.

TIRADENTES, Atalo. **Using Solar Energy to Generate Electricity and Heat Water.** Federal University of Lavras-UFLA LAVRAS MINAS
GENERAL - BRAZIL. MG, 2007;

ZILLES, Roberto et al. **Photovoltaic systems connected to the electricity grid.** São Paulo: Oficina de textos, 2012.

CHAPTER 7

ANNEXES

ANNEX A - General schematic drawing.

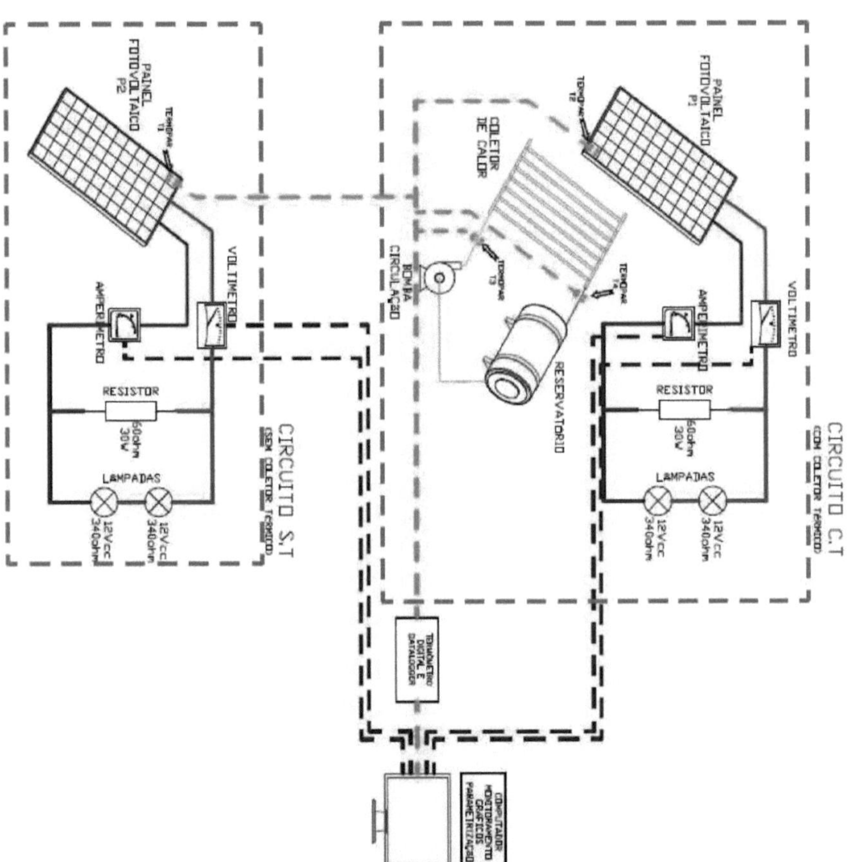

Schematic demonstration of the electrical diagram for both cases, i.e. photovoltaic module P1 with the installation of a heat collector and P2 without a heat collector. It is also possible to visualise the installation of the instruments and temperature sensors.

ANNEX B - Pump used for water circulation

- Bomba de funcionamento submerso ou externo*;
- Eixo especial em alumínio-cerâmica para uso em água doce ou salgada;
- Estrutura dupla de resfriamento para rápida dissipação do calor;
- Resina especial de alta qualidade para isolar o motor do contato com a água;
- Bomba em plástico ABS de alta qualidade, que não altera o pH e é anticorrosivo;
- Fortes ventosas para fixação da bomba e que não ressecam;
- Os modelos 2500 e 4000 acompanham 2 bicos de chafariz com altura e formato ajustáveis. Os modelos 6000 e 9000 acompanham somente o bico Tulipa.

*Exceto modelo 650, apenas de uso submerso.

Curvas de desempenho

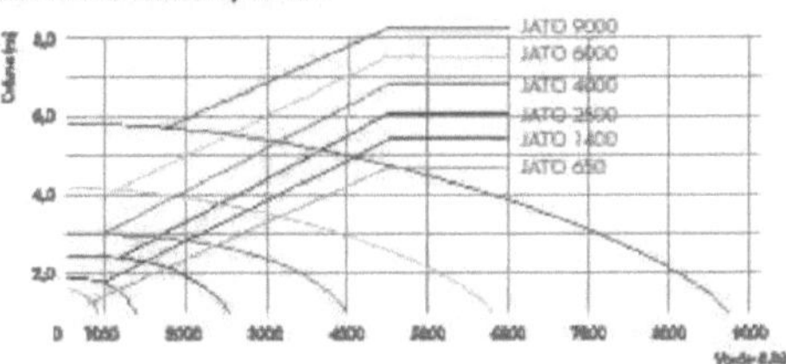

Informações Técnicas

Modelo	Uso externo	Tensão	Frequência	Potência	Vazão	Coluna	Peso	Tamanho (mm)
Jato 650	Não	110-220V	60Hz	9,5W	650 L/h	1,2m	1,22Kg	130x60x85
Jato 1400	Sim	110-220V	60Hz	24W	1400 L/h	1,8m	1,6Kg	138x83x88
Jato 2500	Sim	110-220V	60Hz	44W	2500 L/h	3,35m	1,85Kg	157x85x132
Jato 4000	Sim	110-220V	60Hz	92W	4000 L/h	3,0m	2,4Kg	216x100x140
Jato 6000	Sim	110-220V	60Hz	130W	5800 L/h	4,1m	3,7Kg	242x145x170
Jato 9000	Sim	110-220V	60Hz	200W	8000 L/h	5,0m	3,7Kg	242x146x170

Partes da bomba

(1) Pré-filtro (2) Esponja de filtragem (3) Tampa da câmara d'água (4) Impeller e Rotor
(5) Corpo da bomba (6) Motor (7) Tampa traseira (8) Saída do cabo
(9) Suporte do rotor (10) Ventosas de fixação (11) Anel de vedação

ANNEX C - Ultimeter used to measure current

MULTÍMETRO DIGITAL

Digital Multimeter

ET-1002

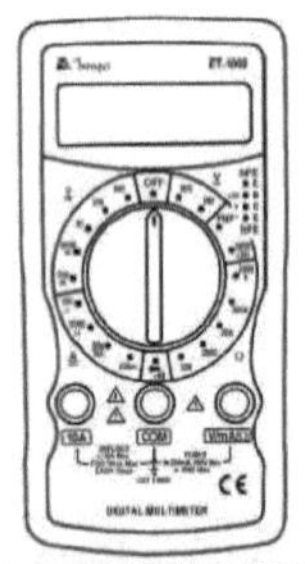

Minipa®

MANUAL DE INSTRUÇÕES

Instructions Manual

Manual de Instrucciones

8) ESPECIFICAÇÕES

A. Especificações Gerais

- Indicação de Sobrefaixa: Dígito mais significativo (1).
- ⚠ Proteção por Fusível para o Terminal de Entrada mA: Fusível de 250mA / 250V. Sem proteção para a entrada 10A.
- Contagem Máxima do Display: 1999.
- Taxa de Amostragem: Aprox. 2~3 vezes por segundo.
- Coeficiente de Temperatura: 0.1 x (precisão especificada) / 1°C, < 18°C ou > 28°C.
- Ambiente: Operação: 0°C a 40°C (32°F a 104°F), RH<75%. Armazenamento: -10°C a 50°C (14°F a 122°F), RH<85%.
- Altitude de Operação: 2000m.
- Tipo de Bateria: 1 x 9V (NEDA1604 ou 6F22 ou 006P).
- Duração da Bateria: Aprox. 200h (típico).
- Indicador de Bateria Fraca: 🔋.
- Segurança / Conformidade: IEC1010 Sobretensão e Dupla Isolação. CAT I 600V.
- Dimensões: 128(A) x 66(L) x 27(P)mm.
- Peso: Aproximadamente 130g (incluindo bateria).

B. Especificações Elétricas

Precisão: ± (a% leitura + b dígitos), garantido por 1 ano.
Temperatura de operação: 23°C ± 5°C, Umidade relativa: < 75%.
As precisões são especificadas de 5% a 100% da faixa ou especificado de outra maneira.

17

A. Tensão DC

Faixa	Precisão	Resolução
200mV	±(0,5%+5D)	100μV
2000mV	±(0,8%+5D)	1mV
20V		10mV
200V		100mV
600V		1V

Observações:
- Impedância de Entrada: 1MΩ.
- Proteção de Sobrecarga: 600V DC / AC RMS.

B. Tensão AC

Faixa	Precisão	Resolução
200V	±(1,2%+10D)	100mV
600V		1V

Observações:
- Impedância de Entrada: 500kΩ.
- Resposta em Freqüência: 40Hz ~ 400Hz.
- A tensão AC é mostrada como o valor eficaz para onda senoidal (RMS).
- Proteção de Sobrecarga: 600V DC / AC RMS.

C. Corrente DC

Faixa	Precisão	Resolução
200μA	±(1,0%+5D)	100nA
2000μA		1μA
20mA		10μA
200mA	±(1,2%+5D)	100μA
10A	±(2,0%+5D)	10mA

Observações:
- Corrente Máxima: 10A (tempo de teste menor que10 segundos para medida na faixa de 10A e com intervalos de 15 minutos entre medidas).

18

- Proteção de Sobrecarga: Fusível de 0,25A / 250V na Entrada mA; Sem Fusível na Entrada 10A.

D. Resistência

Faixa	Precisão	Resolução
200□	±(1,0%+5D)	100m□
2000□	±(0,8%+5D)	1□
20k□		10□
200k□		100□
2000k□	±(1,2%+5D)	1k□

Observações:
- Tensão em Aberto: < 3,2V.
- Na faixa de 200Ω, curto circuite as pontas de prova, e o valor mostrado deve ser subtraído das leituras das medidas para maior precisão.
- Proteção de Sobrecarga: 250V DC / AC RMS.

E. Teste de Diodo / Continuidade

Faixa	Descrição	Condição de Teste
⊣▶⊢	O display exibe a queda de tensão aproximada do diodo.	Corrente direta de aprox. 1mA, e tensão reversa de aprox 2,8V.
◁◁	A buzina toca se a resistência medida for menor que 50□.	Tensão de Circuito Aberto de aprox. 2,8V.

Observações: Proteção de Sobrecarga: 250V DC / AC RMS.

F. Teste de hFE de Transistor

Faixa	Descrição	Condição de Teste
0~1000	O display exibe o valor de hFE do transistor em teste	Corrente de Base de 10μA e Vce de 2,8V.

19

ANNEX D - Multimeter used to measure voltage

impac

www.impac.com.br

Rua Murtinho Nobre, 17 - CEP 05502-050 - Butantã - São Paulo - SP
Tel.:(11)3816-0371 ou (21) 3521-6371 - E-mail: vendas@impac.com.br - RV 12/09

Manual de Instruções

Multímetro Auto Range
Com Interface RS-232

IP-336RS

-Para auxiliar na leitura o programa pode exibir um gráfico e também uma planilha com os dados da medição.(Fig.8)
-Para mostrar a lista de dados clique em "List" e para mostrar o gráfico clique em "Graph" para adicionar mais leituras clique em "Recording" e em "Continue"

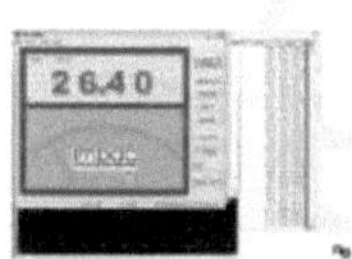

Fig 8

Salvar e imprimir os dados

-Para salvar os dados da medição clique em "File" e em seguida em "Save Data File"(Fig.9)
-Para imprimir clique em "File" e em seguida em "Print" e escolha entre imprimir a Lista (List) e o Gráfico (Graph).(Fig.10)

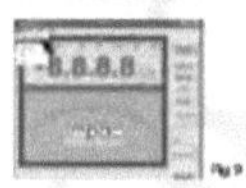

Fig 9

Fig 10

Configuração

-Para editar e configurar as medidas clique em "Option" e e seguida em "Configure".(Fig.11)
-Em seguida uma nova tela será mostrada onde você poderá definir as unidades de medida, os valores maximos e minimos e o tamanho do gráfico.(Fig.12)

Fig 11

Fig 12

9

Características

Mostrador: LCD 3¾ digitos com leitura max. de 4000
Controle das Faixas: Automático / Manual
Polaridade: Polaridade negativa indicada automaticamente no mostrador
Ajuste ZERO: Automático
Indicação de sobre faixa: "OL" é indicado no mostrador
Indicação de bateria fraca: [battery symbol] é indicado no mostrador
Desligamento Automático: Após 30 min sem uso
Temperatura de operação: 0°C ~ 40°C
Umidade de operação: < 85%
Fusível: 0,5 A / 250 V Ø 5 x 20 mm e 10A / 250 V Ø 6 x 25 mm
Alimentação: 2 pilhas de 1,5V AA
Dimensões: 145 x 73 x 40 mm
Peso: 260g
Os medidores são projetados de acordo com a norma IEC-1010 sobre instrumentos de medição eletrônica, classe de proteção (CAT II) 600V e grau de poluição 2

Tensão CC

ESCALAS	RESOLUÇÃO	PRECISÃO
400 mV	0,1 mV	± (0,5% + 5 digitos)
4 V	1 mV	± (0,5% + 5 digitos)
40 V	10 mV	± (0,5% + 5 digitos)
400 V	0,1 V	± (0,5% + 5 digitos)
600 V	1 V	± (0,8% + 5 digitos)

Impedância de entrada: 10MΩ

Tensão CA

ESCALAS	RESOLUÇÃO	PRECISÃO
400 mV	0,1 mV	± (1,2% + 5 digitos)
4 V	1 mV	± (0,8% + 5 digitos)
40 V	10 mV	± (0,8% + 5 digitos)
400 V	0,1 V	± (0,8% + 5 digitos)
600 V	1 V	± (1,2% + 5 digitos)

Impedância de entrada: 10MΩ   Resposta de Frequencia: 40 ~ 400Hz

Continuidade

ESCALA	FUNÇÃO
[diode/buzzer symbol]	A buzina interna soara quando a resistência for menor que 70Ω

Proteção de Sobrecarga: 250V CC/CA RMS

2

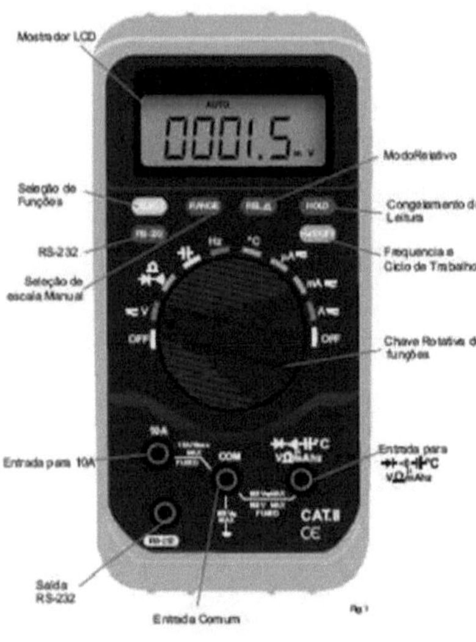

Fig 1

Cuidados com o Equipamento

-Mantenha o IP-336RS sempre seco, se por acaso ele molhar seque-o imediatamente, porque o liquido pode corroer o circuito interno do aparelho.
-Use e guarde o multímetro apenas em locais com temperatura e umidade normais, os extremos de temperatura e umidade podem encurta a vida dos dispositivos eletrônicos, danificar a bateria e distorcer ou derreter as peças plásticas.
-Manuseie o aparelho cuidadosamente, deixá-lo cair pode danificar as placas do circuito e danificar a integridade física do multímetro, embora o "holster" possa fornecer bastante proteção.
-Mantenha o IP-336RS longe da poeira e da sujeira, pois podem desgastar as peças.
-Limpe o multímetro com um pano úmido ocasionalmente, não use produtos químicos como solventes e detergentes.
-Use sempre baterias novas de boa qualidade, retire sempre as baterias velhas e fracas, pois podem liberar resíduos que destroem o circuito eletrônico.

Troca da bateria e do fusível

-Assegure-se que o multímetro esteja desligado e as pontas de prova removidas.
-Remova o parafuso e retire a tampa da caixa da bateria.
-Retire as pilhas velhas e substitua por novas do mesmo tipo.
-Remova os parafusos e levante a caixa. Substitua o fusível por outro com o mesmo tipo e avaliação: 5×20mm 0.5A/250V ou 6×25mm 10A/250V

Garantia

Este produto tem garantia pelo prazo definido no certificado de garantia ou na nota fiscal, a garantia cobre somente defeitos de fabricação, a partir da data da emissão da nota fiscal para o primeiro comprador.

Ocasionará perda da garantia:

1.Comprovação de queda ou utilização de maneira que comprometa os circuitos internos do aparelho.

2.Vestígios de abertura por técnicos não autorizados

3.Quebra física dos sensores

4.A garantia cobre única e exclusivamente defeitos de fabricação. Sobre hipótese alguma a garantia cobrirá defeitos gerados por erro ou mau uso. 10

Corrente CC

ESCALAS	RESOLUÇÃO	PRECISÃO
400 µA	0,1 µA	± (2,0% + 5 dígitos)
4000 µA	1 µA	± (2,0% + 5 dígitos)
40 mA	10 µA	± (1,5% + 5 dígitos)
400 mA	0,1 mA	± (1,5% + 5 dígitos)
4 A	1 mA	± (2,0% + 5 dígitos)
10 A	0,01 A	± (2,0% + 5 dígitos)

Proteção de Sobrecorrente: 0.5A/250V / 10A/250V
ATENÇÃO: Tempo máximo de operação: 15s (4A e 10A).

Corrente CA

ESCALAS	RESOLUÇÃO	PRECISÃO
400 µA	0,1 µA	± (2,5% + 3 dígitos)
4000 µA	1 µA	± (2,5% + 3 dígitos)
40 mA	10 µA	± (2,0% + 5 dígitos)
400 mA	0,1 mA	± (2,0% + 5 dígitos)
4 A	1 mA	± (2,5% + 5 dígitos)
10 A	0,01 A	± (2,5% + 5 dígitos)

Proteção de Sobrecorrente: 0.5A/250V / 10A/250V
Resposta em freqüência: 40 ~ 400Hz
ATENÇÃO: Tempo máximo de operação: 15s (4A e 10A).

Resistência

ESCALAS	RESOLUÇÃO	PRECISÃO
400 Ω	0,1 Ω	± (1,0% + 3 dígitos)
4 kΩ	1 Ω	± (1,0% + 3 dígitos)
40 kΩ	10 Ω	± (1,0% + 5 dígitos)
400 kΩ	100 Ω	± (1,0% + 5 dígitos)
4 MΩ	1 KΩ	± (1,0% + 5 dígitos)
40 MΩ	10 KΩ	± (2,0% + 5 dígitos)

Proteção de Sobrecarga: 250V DC/AC RMS

Temperatura

ESCALAS	RESOLUÇÃO	PRECISÃO
-50 - +200 °C	0,1 °C	± (0,75%± 3°C)*
200 - 750 °C	0,1 °C	± (1,5%± 3°C)*

*Precisão somente do aparelho, adicionar erro do termopar

3

RS-232

Instalação

-Este programa é compatível com windows XP e Vista.
-Coloque o CD do software que vem junto com o aparelho no compartimento de CD do seu computador.
-Abra a pasta do CD e clique duas vezes na pasta "Setup" e em seguida no ícone "SETUP"
-A tela de instalação do programa irá abrir e em seguida clique em "next"(Fig.2)
-O programa será instalado no seu computador.(Fig.3)
-Clique em "Finish" então a instalação do Programa estará concluída.(Fig.4)

Fig 2 Fig 3 Fig 4

Operação

-Conecte uma extremidade do cabo na saída RS-232 do IP-336RS e a outra na porta serial do seu computador (Fig.5) e pressione o botão "RS-232" do aparelho.(O Mostrador do multímetro indicará "RS-232")

Fig 5

-Inicie o programa no seu computador
-Clique em "Begin" para iniciar.(Fig.6)
-Logo em seguida o programa mostrará a medição que o IP-336RS está medindo.(Fig.7)

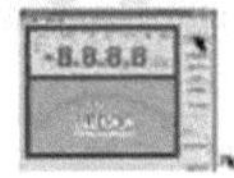

Fig 6

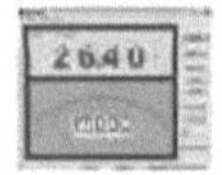

Fig 7

8

ANNEX E - Digital thermometer used to measure temperatures

Termômetro Datalogger

ITTH-1400

Manual de Instruções

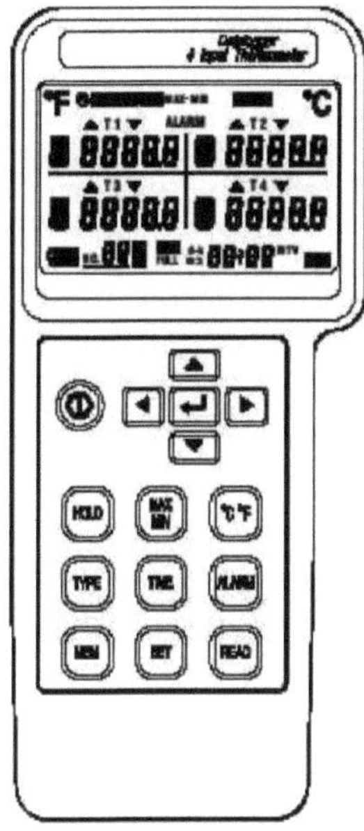

※ Incluso CD com software.

1. Introdução as funções do aparelho

Este instrumento é um termômetro digital com 4 entradas e datalogger, e trabalha com termopares tipo K, J, E, T, R, S, N, L, U, B e tipo C como sensores de temperatura.
Seus indicadores de temperature seguem a escala internacional de 1990. (ITS-90)

- Leia cuidadosamente as instruções de segurança antes de iniciar o uso do aparelho.
- Utilize este aparelho somente de acordo com as condições especificadas neste manual, caso contrario suas condições de proteção e segurança podem ser prejudicadas.

Condições ambiente

① Altitude até 2.000 metros

② Umidade relativa máxima de 80%

③ Temperatura ambiente 0~50°C (32°F ~122°F)

Funções

- Proteção de isolamento de 350Vp-p entre duas entradas.
- Termômetro de alta precisão com termopares tipo K, J, E, T, R, S, N, L, U, B, C.
- Display de temperatura e 4 funções de entrada T1/T2/T3/T4.
- Alarme programável Hi – Lo alarme para 4 entradas.
- Display para valores MAX, MIN e MAX–MIN para as 4 entradas.
- Ajuste independente para entradas (tipo de termopar, alarme de valores Hi – Lo).
- Funções memória e leitura (99 ajustes)
- Capacidade de armazenamento de 512KB.
- Interface USB.

Símbolos de segurança

Utilize somente acessórios de acordo com especificação.

CE Conformidade com EMC.

2. ESPECIFICAÇÕES

2-1 Especificações

Tipo	°C		°F	
	Faixa	Precisão	Faixa	Precisão
K	-200 ~ -150 -150 ~ -100 -100 ~ 999.9 1000 ~ 1370	±3.0°C ±2.0°C ±0.05% ±1.0°C ±0.2% ±1.0°C	-328 ~ -238 -238 ~ -148 -148 ~ 999.9 1000 ~ 2498	±5.4°F ±3.6°F ±0.05% ±1.8°F ±0.2% ±1.8°F
J	-200 ~ -100 -100 ~ 100 100 ~ 999.9	±2.5°C ±1.5°C ±0.05% ±1.0°C	-328 ~ -148 -148 ~ 212 212 ~ 999.9 1000 ~ 1832	±4.5°F ±2.7°F ±0.05% ±1.8°F ±0.2% ±1°F
E	-150 ~ -100 -100 ~ 760	±3.0°C ±0.05% ±1.0°C	-238 ~ -148 -148 ~ 999.9 1000 ~ 1400	±5.4°F ±0.05% ±1.8°F ±0.2% ±1°F
T	-200 ~ -150 -150 ~ -100 -100 ~ 400	±3.0°C ±0.15% ±2.5°C ±0.1% ±1.0°C	-328 ~ -238 -238 ~ -148 -148 ~ 752	±5.4°F ±0.15% ±4.5°F ±0.1% ±1.8°F
R/S	0 ~ 100 100 ~ 300 300 ~ 999.9 1000 ~ 1600	±5.0°C ±3.0°C ±0.05% ±2.0°C ±0.25% ±2.0°C	32 ~ 212 212 ~ 572 572 ~ 999.9 1000 ~ 2912	±9.0°F ±5.4°F ±0.05% ±3.6°F ±0.25% ±3.6°F
N	0 ~ 999.9 1000 ~ 1300	±0.1% ±1.0°C ±0.2% ±1.0°C	32 ~ 999.9 1000 ~ 2372	±0.1% ±1.8°F ±0.2% ±1.8°F
L	-200 ~ 900	±0.1% ±1.0°C	-328 ~ 999.9 1000 ~ 1652	±0.1 % ±1.8°F ±0.2% ±1°F
U	0 ~ 600	±0.1% ±1.0°C	32 ~ 999.9 1000 ~ 1112	±0.1% ±1.8°F ±0.2% ±1°F
B	600 ~ 999.9 1000 ~ 1760	±0.05% ±2.0°C ±0.1% ±2.0°C	1112 ~ 1831 1832 ~ 3200	±0.05% ±3.6°F ±0.1% ±3.6°F
C	0 ~ 999.9 1000 ~ 1760	±0.1% ±1.5°C ±0.2% ±1.5°C	32 ~ 999.9 1000 ~ 3200	±0.1% ±2.7°F ±0.2% ±2.7°F

NOTA

As informações básicas de precisão de temperatura não incluem erros que possam ocorrer por erro nas sondas. Por favor consulte as especificações de precisão das sondas de temperatura para detalhes adicionais.

Coeficiente de temperatura

0.01% leitura +0.1°C por °C (0.2°F por °F)

Printed by Books on Demand GmbH, Norderstedt / Germany